Beneath the Night

Stuart Clark is an astronomer and award-winning science writer for the *Guardian*, *New Scientist*, *BBC Focus*, and many other publications. He is writer and presenter of the 'Music of the Spheres' series for BBC R3, and author of several works of non-fiction and fiction that have been translated into more than twenty-five languages. He is a Visiting Fellow at the University of Hertfordshire, a Fellow of the Royal Astronomical Society, former Vice Chair of the Association of British Science Writers and a consultant for the European Space Agency. In September 2020, the University of Hertfordshire awarded him an honorary Doctor of Science degree for services to astronomy and the public understanding of science.

@DrStuClark

STUART CLARK

Beneath the Night

How the stars have shaped
the history of humankind

First published by Guardian Faber in 2020
Guardian Faber is an imprint of Faber & Faber Ltd,
Bloomsbury House, 74–77 Great Russell Street,
London WC1B 3DA

Guardian is a registered trade mark of
Guardian News & Media Ltd,
Kings Place, 90 York Way, London N1 9GU

This paperback edition published in 2021

Typeset by Paul Baillie-Lane
Illustrations by Chris West Graphic Design

Printed and bound by CPI Group (UK) Ltd, Croydon, CR0 4YY

A CIP record for this book
is available from the British Library

ISBN 978–1–78335–154–1

MIX
Paper | Supporting
responsible forestry
FSC® C013604

4 6 8 10 9 7 5 3

For anyone who has stood beneath the night sky
and wondered about the stars

Contents

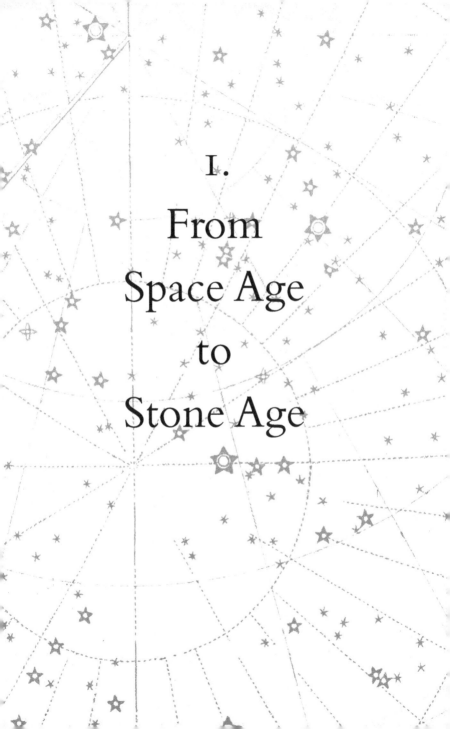

I.

From
Space Age
to
Stone Age

The next time it's a cloudless night, step outside and find somewhere dark. It should be well away from streetlights, trees and buildings so that you can get the best possible view of the night sky – but don't raise your gaze immediately. It will take thirty or forty minutes for your eyes to fully adjust to the dark. During that time, they will become from ten thousand to a million times more sensitive to light. Perfect for stargazing!

Once the time has elapsed and you have found your spot, look up. Depending on the atmospheric conditions and the sensitivity of your eyes, you will be able to see some three or four thousand stars, each one a distant sun in its own right. Each one a possible home to a family of planets.

In most people, this experience tends to bring about feelings of tranquillity and reverence, and often a sense of their own insignificance. Even having spent my entire life studying the night sky, the sight of it never fails to fill me with awe and excitement. As I've sought to understand the stars and come to terms with the sheer vastness of it all, I've lately come to realise that it is not the number or the nature of the stars that is the most enchanting thing about them. It is that, compared to our brief lives, the stars are immortal.

Shakespeare saw the same stars in the same patterns that we do. So did Galileo, Columbus, Joan of Arc, Cleopatra, and the first man-ape to look up in curiosity. From space age back to stone age, to be beneath the night is to witness something that every other human who has ever lived has also seen. It is our common heritage.

This book is a history of our relationship with the night sky. More than a book about our understanding of astronomy, it is the story of how our fascination with the heavens has shaped society, culture and religion as well as science. Beyond enabling our scientific understanding of the universe, the stars have inspired our poets, artists and philosophers; given us a place to project our hopes and fears; revealed our true origin and hinted at our ultimate fate.

The very fact that we look to the night sky in our search for meaning is one of the indelible hallmarks of our humanity. As this book will show, to tell the story of this nocturnal fascination is to tell the story of what it is to be human.

There is no definitive theory as to how or why humans began to relate to the night sky. But there is a growing body of evidence from a number of different disciplines that suggests it is at least plausible to believe our fascination with it started almost as soon as recognisably modern humans evolved, almost seventy thousand years ago.

The modern impetus to look this far back in prehistory comes from the work of an American journalist turned archaeologist called Alexander Marshack. As with much of the rest of the world at the time, Marshack's fascination with space began on 4 October 1957, when the Soviet Union succeeded in making a rocket powerful enough to launch the world's first spacecraft, Sputnik One. What makes Marshack stand out from many of his contemporaries, however, is that he wasn't just fascinated by the technological achievements of the Space Age. His interest was primal: he wanted to know what had driven humans to want to 'touch' the night sky.

Five years later, in the autumn of 1962, President John F. Kennedy made his now famous speech at Rice Stadium in Houston, Texas, in which he committed the USA to landing a human on the Moon's surface before the decade was out. Marshack set out to write a book similar to that which you're reading now, in an attempt to explain how – and, crucially, *why* – humankind had reached the point in history where it was possible to bring such a mission into reality. But as soon as Marshack began his research, he found it 'an almost impossible task'.[1]

He spent much of 1963 travelling across America interviewing people involved in the burgeoning field of space exploration. His interviewees included many of the pre-eminent experts of the time, such as President Kennedy's science adviser Dr Jerome Wiesner, NASA's chief administrator James Webb, representatives of the National Academy of Sciences and the Air Force, and numerous academics. He also spoke to their counterparts in the Soviet Union. But no one could give him a definitive answer to the simple question of *why* humankind was exploring space. It was as if the urge to do so was a basic human compulsion.

And, indeed, examples of this particular compulsion resonate through history. In 1596, the great German mathematician and astronomer Johannes Kepler wrote:

We do not ask for what useful purpose the birds do sing, for song is their pleasure since they were created for singing.

Similarly, we ought not to ask why the human mind troubles to fathom the secrets of the heavens ... The diversity of the phenomena of Nature is so great, and the treasures hidden in the heavens so rich, precisely in order that the human mind shall never be lacking in fresh nourishment.[2]

Even further back, around 2,400 years ago, the Greek philosopher Plato wrote his classic work *The Republic*. In Book VII, he hypothesised that our eyes were formed for the study of the night sky, but that rather than letting its sheer beauty beguile us, we should exercise our minds to understand the order behind the celestial arrangements. Again, Plato's implication is clear. The reason we should study the night sky is the same one that British explorer George Mallory gave when asked why he wanted to climb Mount Everest: 'Because it's there.' Kennedy even used Mallory's quote in Houston for the reason why America should land on the Moon.

★   ★   ★

To try to explain the emotional draw of the night sky, Marshack first tried to identify when our fascination with it began. This search took him back to the time before civilisation and agriculture, before history itself, to when humans lived in hunter-gatherer communities tens of thousands of years ago. Instead of a book about space, he ended up writing one about the prehistoric origin of human science and culture – and the pivotal role that the night sky played in our awakening. As his wife was quoted as saying in his 2004

New York Times obituary: 'He was so intrigued that he left the space age and went back to the ice age.'

The ice age in question was the one that gripped the world from 2.6 million years ago to just twelve thousand years ago. During that time, most of northern Europe was submerged beneath the arctic ice sheets, and the glaciers of the Alps reached far beyond their modern boundaries. It was also during this period that various species of humans emerged as distinct from the other great apes. This process began in Africa around 2.3 million years ago with the appearance of *Homo habilis*, and culminated approximately 200,000 years ago with the arrival of our own species, *Homo sapiens*. However, the tipping point for our story did not come until another 130,000 years after the arrival of our species, when something truly special happened: we began to think differently.

No one knows why this should have happened. It could have been some random mutation in our DNA that suddenly allowed our brains to perceive the world in more abstract ways, or it could have been a gradual process that began much earlier with the appearance of *Homo sapiens*.[3] Whatever the trigger, by seventy thousand years ago, the so-called human revolution was complete.[4] And despite the tens of thousands of years that have since elapsed, it is thought that there is no essential difference between current humans and our ancestors from that period. Their brain power was the same as ours, their ability to reason was the same as ours, and so too their curiosity and capacity to dream. All these primitive humans lacked was the knowledge that we have now accumulated. But the fossil record shows that they were learning fast.

By around forty thousand years ago, a human population of some five million (as compared to today's eight billion) had spread from Africa across the globe. Archaeologists identify this period as the Upper Paleolithic. It extends from around fifty thousand to ten thousand years ago. As hunter-gatherers, the humans of this era obtained their food by the collection of wild plants and the trapping of wild animals. In the artefacts left behind we can see the development of logical thought that leads to technology: oil-lamps, boats, bows and arrows, sewing needles. And there are more than tools on offer.

Art was also born in this period. The earliest undisputed pieces that display creative thinking date to around forty thousand years ago and were found in the Hohle Fels ('hollow rock') cave near Schelklingen, in the Swabian Jura region of Germany. They include figurines such as the Venus of Hohle Fels and a 'flute' made from a hollow vulture bone. In the nearby Stadel cave, a similarly ancient ivory figurine of a lion was found. Carved from the tusk of a woolly mammoth, what sets it apart is that the lion is standing on its hind legs in a human pose. So, the lion-man of Stadel suggests that the artist had an imagination that could conceive of things that do not exist in reality – in this case a lion-human hybrid.

But what captured Alexander Marshack's imagination was a ten-centimetre-long piece of fossilised baboon bone that had been found in the ruins of Ishango, an ancient Congolese village on the shores of Lake Edward. Unearthed in 1960 by Belgian archaeologist Jean de Heinzelin de Braucourt, it was around twenty thousand years old and notable because it had been carved in a rather unaesthetic way with a multitude of notches. Although they are hardly a work of art, neither do they appear to be random. They are grouped into three distinct regions. The first contains sub-groupings that hold 11, 13, 17 and 19 lines; the second grouping reads 3, 6, 4, 8, 10, 5, 5 and 7; while the third reads 11, 21, 19 and 9.[5]

Describing the find in *Scientific American*, de Heinzelin pointed out that the first group are prime numbers between 10 and 20.[6] The third group represent a mathematical pattern: 10+1, 20+1, 20-1 and 10-1. But the second group defied his ability to find order. Despite this failure, he speculated that it could have been carved by someone who was playing some sort of arithmetical game. Marshack balked at this interpretation. To him, the scratches looked more like tally marks – but tallies of what?

Marshack remembered a paper he had read about modern hunter-gatherer societies, such as the Kalahari bushmen of Africa. In that work, the authors described how those societies have some knowledge of reckoning the passage of time by using the stars and/or the Moon.

The night sky is perfect for this purpose. Firstly, the days are related to the Sun and the way its movement heralds day and night. The year and its seasons are clearly related to the stars and the way the constellations change their position over a twelve-month period. The month, in its simplest form, refers to the time it takes for the Moon to complete a four-week cycle of its phases. The waxing phase from new moon to full moon takes about fourteen days, punctuated one week in by a half moon. The same, in reverse, is true of the waning phase.

The similarity of the words month and Moon is no coincidence, either. Although the etymology of the words is complex, they do share an origin in the Latin word *metiri* (to measure). This would suggest that the Moon has been well established as a yardstick for the passing of time for more than two thousand years.

Marshack wondered if this usage extended back into the Upper Paleolithic. Specifically, he wondered if the Ishango bone was a tally of the phases of the Moon. If so, that would make it the world's oldest known calendar and would suggest that the earliest known relationship that humankind had to the night sky was one of practicality: they used it as a clock.

It would also mean that we began our relationship with the night sky almost as soon as we possibly could: during the great human revolution when our ancestors first thought about the world around them, about how to live in it and the meaning of their place in it.

Marshack set to work to test his hypothesis and came up with a convoluted system that did indeed appear to correlate the scratch marks to the phases of the Moon. But to do so meant assuming that whoever carved the bone had grouped the lunar observations into two sequences of sixty days, and one of forty-eight days, when there was no clear reason why someone would do this. As a result, while his interpretation of the Ishango bone is a compelling idea, it could hardly be seen as conclusive. Indeed, since Marshack other researchers have suggested alternative interpretations that range from the extraordinary (a stone age 'slide rule') to the mundane (a tally of goods).

Seeking further evidence for his theory, Marshack sought out other similarly notched artefacts from the Upper Paleolithic period, eventually publishing his findings in *The Roots of Civilization* in 1972. Though his work was controversial, with a common criticism arguing that it was too speculative, he has nonetheless been an inspiration to subsequent researchers, who continue to look at artefacts and other markings for possible astronomical interpretations. And while it is clear that proving this point from the archaeological data alone is a difficult task, the general feeling persists that Marshack's theories do have some merit, and other artefacts found since have only added further weight to his argument.

One of these is an elephant's tibia found at the prehistoric site of Bilzingsleben, in Thuringia, Germany. It had been carved with a total of twenty-one parallel lines in two groupings. One grouping contains seven lines, the other fourteen, but the bone is broken. The palaeontologists who found it, Dietrich and Ursula Mania, proposed that the missing piece could contain a mirror of the first group of marks, bringing the total number to twenty-eight, a number that immediately reminds us of the lunar month. If so, the bone could mark the seven days from new moon to waxing half moon, then the fortnight through full moon to waning half moon, and the final seven days that lead back to new moon. Even though this interpretation is highly speculative, such an artefact would not usually stand out from any of the other putative lunar calendars if it were not for the fact that it is much older. Rather than tens of thousands of years, the elephant bone has been dated to between 350,000 and 250,000 years old.[7] Mind-blowingly, this places it before the human revolution, before even the evolution of *Homo sapiens* and back into the time of *Homo erectus*, an earlier species of human.

While by no means conclusive, the Bilzingsleben bone and the artefacts studied by Marshack certainly offer tantalising evidence for the notion that Palaeolithic people kept track of the night sky. But accepting this leads to a larger mystery: why? What motivated these early humans to do this?

The various answers to this question suggested by scholars over the decades usually fall into one of two categories: practical or religious. According to the practical school of thought, the night sky was studied because it could be used to mark the passage of time. At the opposite end of the spectrum, the religious theorists postulate that the emotion of awe we feel when we look at the night sky transforms into a need to worship it. So we study the various movements of the Sun, the Moon and the other celestial objects to venerate them as gods.

However, neither suggestion really works: both impose a false dichotomy between religious and practical motivations that fails to capture the broad range of human thinking. Remember that those early *Homo sapiens* had the same brain power that we do. Their minds were capable of every emotion and desire that flows through us today.

So let's reframe the question. Considerable effort is needed to observe and painstakingly record the night sky for nights, weeks, months, even years on end. This is true today, and would have been truer still in a hunter-gatherer society where free time was at a premium. There must, therefore, have been some strong *societal* advantage to doing so. What might that have been?

For an answer, we can turn to the modern-day hunter-gatherer societies and the work of the ethnographers. Ethnography is the observation of a society's culture. Since it is impossible to travel back in time to the Upper Paleolithic and observe the hunter-gatherer tribes that roamed the Earth, the next best thing is to observe those who still live like this today. If these modern hunter-gatherer societies use astronomical knowledge for societal benefit, it would provide a compelling argument in favour of the theory that so too did the hunter-gatherers of the Upper Paleolithic.

There are an estimated one hundred uncontacted tribes in the world, mostly found in Amazonia and New Guinea.[8] They mostly manage to avoid contact with the outside world and often meet any encroachment with hostile force. So ethnographers are forced to choose others who are more receptive to contact, yet have shunned the trappings of the modern world. Of these, there are many dozens.

Next, ethnographers split hunter-gatherers into two sub-groups: simple and complex. Simple hunter-gatherer groups are those with low population densities. They are completely egalitarian, with no social hierarchy and with all resources being completely shared. Their counting systems do not extend beyond a few tens.

Complex hunter-gatherer groups tend to arise when the density of people increases. In these societies, there is an emerging hierarchy, usually to do with surplus of food; those families who produce the most have a higher status than the others. There is also a tendency in these groups for families to own small patches of land, and to trade both food and primitive *objets d'art*. In trading or loaning food and other items, keeping a tally of the debts is clearly important. This leads to record-keeping and complex counting systems that extend into the hundreds and thousands.

As noted by Marshack, almost all extant hunter-gatherer groups have some kind of astronomical system to help them reckon the passage of time, but there is a fascinating division between the two types of group.

The simple hunter-gatherer groups have a knowledge of the phases of the Moon and solar events such as solstices, yet they do not go to the trouble of organising feasts, rituals or celebrations around these events. This is entirely in keeping with the fact that they are basically in a fight to stay alive and seldom have the food surpluses that are necessary to make ritual feasting possible.

The picture is quite different for the complex hunter-gatherers. Here, most tribes observe the solstices in some way and maintain some form of lunar calendar, or at least monitor the phases of the Moon. In terms of the solstices, it is the winter solstice – i.e. the shortest day of the year – that appears to be the most important for the group as a whole. It is used to mark the beginning of a period of celebration and feasting – the winter ceremonial. Here, the surplus food for feasting comes from the wealthier families and is used as a way to gather allies and increase their importance among the tribe.

Perhaps the most important fact is that the winter ceremonial is usually presided over by a shaman, an elder, or some other individual that the group recognises as possessing special knowledge relating to the night sky. This person is usually associated with the dominant family of the tribe and is responsible for predicting the coming winter solstice and other astronomical alignments. As such, they are therefore responsible for setting the date of the various feasts and rituals that punctuate the group's year.

To hunter-gatherers these celebrations aren't mere social events; they have a distinctly political dimension. In the same way we use elections to choose our leaders, the various families in the tribe jockey for power at these get-togethers, identifying who can share the most and using this as a means of displaying their wealth. Alliances are brokered, debts repaid, new loans made. They set the agenda and political landscape for the coming year.

A person who saw this at first hand was Canadian anthropologist Thomas Forsyth McIlwraith. Between 1922 and 1924, he spent extended periods of time living with the indigenous Nuxalk people of the Bella Coola valley in British Columbia.[9] He provided a detailed account of their winter ceremonial and the keen way its precise date was debated by the specialists, often leading to bitter disputes. Calculating the solstice dates requires a knowledge of astronomy that only the most powerful families can spare individuals to learn, and so for a lesser Nuxalk family to be able to show that the leading family's astronomer has made an error in their calculations is considered a coup indeed.

By backtracking from these petty human arguments today, we arrive at a plausible reason for the development of accurate monitoring of the celestial realm in Upper Paleolithic times: it was all about jockeying for status on Earth. This argument was proposed in 2011 by Brian Hayden and Suzanne Villeneuve of Simon Fraser University, in British Columbia, Canada, in their paper *Astronomy in the Upper Paleolithic?*, and offers a powerful (and undeniably human) answer to the question of *why* we study the night sky.[10]

To accept the socio-political reason for our astronomical interest, all we have to do is accept that such ceremonials have been a facet of hunter-gatherer societies throughout human existence. In support of this is another aspect of modern hunter-gatherer behaviour that is directly reminiscent of Upper Paleolithic sites around the world: the use of caves as sacred astronomical places.

★   ★   ★

In September 1940, French teenager Marcel Ravidat was exploring some woods near the village of Montignac, in the Dordogne region of south-west France, when he discovered the entrance to a set of prehistoric caves that would become a world sensation and keep archaeologists busy to this day. The caves are those of Lascaux, and they have been designated a UNESCO World Heritage Site for good reason.

At the end of a long entrance shaft, they open into a number of chambers that are covered in striking paintings of animals. After painstaking investigation, archaeologists have deduced that these paintings were produced by a combined effort over many generations that date to around seventeen thousand years ago.

They are by no means unique. Cave art is found across the world, and often dates back tens of thousands of years to the human revolution. Depictions of animals tend to dominate, as do hand stencils. The latter were created by an individual placing their hand on the cave wall and blowing a pigment across it, to create a kind of silhouette on the rock. Intriguingly, while both adult and children's hand stencils are a common feature, the caves themselves do not usually contain artefacts relating to ongoing habitation. So they were not homes in which families lived but places that people visited for some particular reason or another.

At the turn of the millennium, the independent researcher Chantal Jeguès-Wolkiewiez proposed that the Lascaux caves were not randomly chosen as the location for the art work they contain. She showed that Lascaux and similar caves at Bernifal, France, are penetrated by the Sun's rays at sunset on just one single day of the year: the summer solstice.[11] Perhaps, Jeguès-Wolkiewiez argued, this was when the cave was visited, in order to perform a special or sacred ceremony. Ethnographic studies of modern hunter-gatherers lend weight to this interpretation.

The calendar experts of the modern Chumash of southern California have formed an élite society. Known as the 'antap, they both preserve astronomical knowledge and guard it from the common folk. Members of the 'antap ritually pass on their secret knowledge to specially chosen young initiates in elaborate ceremonies that take place in special caves, decorated with rock art. These caves are also ones in which the Sun penetrates the entrance only on the summer solstice. And the Chumash are not alone in doing this.

In a cave in Wyoming, Crow Native Americans have painted the image of a bison on a wall that is illuminated only at the winter solstice, when the Crow begin their prayers for a successful bison hunt season. Indeed, it seems that some form of élite or even secret society that revolves around calendrical and astronomical knowledge is a feature of most complex hunter-gatherer societies.

Membership of these secret societies relies on an individual being powerful and well connected in the community. It also relies on being able to buy your way in. So, gradually the society gains wealth and power in the community. Then, at a certain point, something truly fascinating takes place. The society begins to surround what it does in myth, further aggrandising its role and importance to the community it started out supporting. As the myths grow, the roles reverse until the community ends up working to carry this élite. In proposing this scenario, Hayden and Villeneuve termed these élite individuals 'aggrandizers'. They wrote: 'Aggrandizers are individuals who systematically seek to promote their own self-interests above those of other community members and typically employ a variety of subterfuges or strategies in order to achieve these ends.'[12]

And it all starts when a hunter-gatherer community begins to be successful enough to start producing surpluses. This allows effort to be put into other things than simply staying alive, and some individuals see it as a chance for their own advancement.

While that may at first sound a cynical explanation for the underlying reason behind our historical observations of the night sky, it also describes *the* crucial moment in our relationship with it. Namely, it is when we first project our enchantment onto the stars. By endowing the night sky with special mystical properties, early humans marked it out as a different realm from that found on Earth and that set the stage for the ultimate mythologising of the night sky that was to come: religion.

2.

The

Invention

of

Heaven

We find places of worship across the world: mosques, churches, synagogues, temples. Often they form the nucleus of a community, both socially and geographically. They are places where people come together to celebrate a set of shared beliefs that give meaning to the world around them and the lives they lead within it.

Religious beliefs are always anchored to some kind of 'hidden realm', something we cannot directly see but which promises that an underlying order permeates the often random events we endure. In this way, religion strengthens the bond between the various members of the community by providing a set of common values and an underlying framework that gives order to their lives.

In the world today, more than four thousand religions are practised. Many of them are complex edifices of belief that require expert interpretation, but the first religions were much simpler affairs. Like animals and plants, they have evolved into more complex forms or passed into extinction. American philosopher Daniel Dennett estimates that hundreds of thousands of religions have been practised at some time or another on Earth.

The simplest religions can be found in the surviving hunter-gatherer groups. These tend to constitute a form of respect and worship for the natural world termed 'animism' by nineteenth-century anthropologists and philosophers. The term comes from the Latin word *anima* meaning breath, spirit or life, and although it has waxed and waned in popularity since, it addresses how indigenous people relate to nature.

Animism revolves around the belief that 'spirits' pervade everything that exists, from humans and animals to plants and inanimate objects like rocks. When a tree is cut down for building material, or an animal killed for food, these spirits or 'souls' are lost. So, animists believe in the need to be compassionate and respectful towards all natural objects.

Animism is thought to have dominated our ancestors' thinking. It was derived from asking questions about what being alive really means. What is the difference between a human, an animal, a plant and a rock? How should we relate to these other natural objects?

The conclusion, according to animism, is that all of nature is connected via 'spirits'. The individual souls of the smallest drops of water and the furthest stars are all interconnected to form a single universal entity. Disturbances or events in one location ripple through to all other parts – and for 'proof' of this theory they turned to the heavens.

We now know that the Earth rotates around its axis every twenty-four hours, giving us day and night. We also know that our planet orbits the Sun every year. Because Earth's rotation axis is inclined at 23.5° instead of being upright, this gives us the seasons. But our ancestors thought Earth was the fixed centre of the cosmos, and that all the celestial objects revolved around us. From this perspective, all they saw were shifting patterns of stars and planets and simultaneous changes on Earth. And the archaeological evidence shows that they set their minds to studying it.

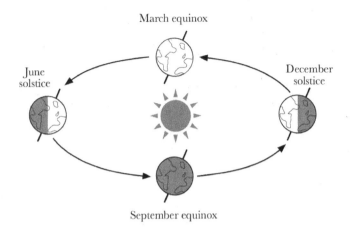

March equinox

June
solstice

December
solstice

September equinox

The seasons are created by the Earth's location in its orbit, which determines the direction of the Earth's rotation axis in relation to the Sun.

★ ★ ★

Some time between 9000 and 7300 BCE, a nomadic group of humans chanced upon a lake in what is now the Nubian desert in the eastern Sahara. Called Nabta Playa, its climate back then was very different from today. The lake formed temporarily each year after the summer rains, and the surrounding plain sprang into fertile life. Judging by the archaeological remains, the wandering humans herded cattle, making them pastoralists rather than hunter-gatherers. So the attraction of Nabta Playa as a place to rest and let their cattle graze is obvious. The seasonal lake could even have become an annual gathering ground for various tribes and groups to meet.

Artefacts make it clear that the first group to find Nabta Playa was not the last. As centuries and then millennia marched onwards, more and more wandering groups of people converged here and, by 7000 BCE, a growing number of them had settled.[13]

Deep wells, surrounded by villages consisting of small huts built in straight lines, indicate that people were living there all year round. The organic remains in the soil show that the inhabitants were eating a variety of wild plants including grains, legumes, tubers and fruits. A few centuries later, goats and sheep appear in the archaeological record. Clearly, the place was orderly and the community followed rules, otherwise it could never have grown. In short, civilisation was beginning. Then something disastrous happened.

Around 7,500 years ago, severe droughts hit the area, the lake dwindled, desert began to encroach and the region was abandoned. When the rains returned a millennium later, so too did humans. But they were markedly different from their predecessors. They practised elaborate rituals such as sacrificing cows and burying them in clay-lined chambers, which they marked with stone-covered mounds. This is significant because the killing of an animal that could have offered the group food is clearly something that cannot be undertaken lightly. Even today, African pastoralists revere their cattle. As the providers of milk, cattle are only slaughtered to mark major turning points in the life of the community. The practice is termed the 'African Cattle Complex'. So, back on the Nabta Playa, the act of animal sacrifice must have provided something that seemed higher than basic day-to-day survival. In other words, it was a religious act.

Along with the sacrifices, the inhabitants erected great megalithic stones arranged in five lines around a central point, like spokes on a wheel. They also built a four-metre-wide stone circle consisting of almost forty stones. Some were flat and almost buried, others stood upright like gnarled teeth sticking out of the desert.

The Nabta Playa site was serendipitously discovered by American anthropologist Fred Wendorf in 1974. It is said that during a long and arduous crossing of the desert, Wendorf and colleagues stopped at random for a comfort break. In the midst of relieving themselves, they noticed shards of pottery and other artefacts strewn across the sand.

Over the course of the next decade, the academics returned time and again, excavating the site and building up a picture of life at Nabta Playa. Key to that picture was their interpretation of the area as a regional ceremonial centre. The bones of the sacrificed cattle were a cornerstone to this view, as was the discovery of the stone circle, which Wendorf referred to as a calendar circle. He was forced to imagine its original appearance, because many of the blocks were broken or had toppled over, yet he was certain enough in his reconstruction to claim that the stones formed a number of astronomical alignments.

He pointed to four pairs of larger stones in the circle, and called the narrow gap between each pair a gate. He showed that the line through two of these gates pointed 'generally north–south', and that the gates of the other two pairs pointed some 70° east of north, which he identified as the calculated point of sunrise on Midsummer's Day six thousand years ago.[14] This combined with the nearby discovery of the cattle sacrifices suggests an ancient link between religious practices and the observation of the stars.

The Nabta Playa stone circle is one of the oldest in the world, and during the next few thousand years, the urge to build such places spread at an incredible rate. There are many thousands of such circles – also known as henges – that remain today. They are made of earth, wood and stone, differ considerably in size, and were certainly constructed for a number of different reasons. Yet the ones that tend to capture our attention the most are the ones like Nabta Playa that appear to be aligned with celestial objects or events. Of these, the most famous is Stonehenge in the UK.[15]

Situated in the county of Wiltshire, Stonehenge is an extraordinary place by any measure. It has become an icon of Stone Age culture with around a million visitors each year. The stone monument visible today dates to around 2500 BCE, but the site itself has a much longer history stretching back to around 8000 BCE – roughly the same time that the Nubians were discovering the lake at Nabta Playa. Archaeological remains show that a community lived at Blick Mead, a spring about one mile away from the site. The inhabitants of Blick Mead are likely responsible for building the first monument, which was an alignment of three large timber posts, each around 0.75 metres in diameter. They placed them close to where the famous stone circle now stands and aligned them east–west. Given the nightly cycle of stars rising in the east and setting in the west, it seems highly probable that these people attached meaning to the sky.

Around 4000 BCE, the people of the area began to construct concentric circular ditches known as causeway enclosures. These ditches were found to contain pottery and human remains, suggesting that they were cemeteries.

Other types of communal burial mounds, known as long barrows, have also been found in the vicinity and date from this period.

Approximately a thousand years on, the first work began at the central site. It consisted of a circular bank and ditch, within which fifty-six one-metre-wide chalk pits were dug in a similar circular pattern. The cremated remains of sixty-three individuals have been found in these pits, and there is evidence that standing stones were used as grave markers. These stones remain at the site but are no longer marking the chalk pit graves. They are called bluestones and are a distinctive form of rock transported 150 miles from the Preseli Hills in North Wales to be used at Stonehenge. The circle formed when they were used as grave markers and was the first stone monument at the site. There were probably timber structures in the centre of the circle as well, though the exact nature of these is unclear from the archaeological remains.

Then, around 2500 BCE, the Stonehenge we know began to take shape. The giant sarson stones that form the iconic stone circle arrived. It is possible that they were collected opportunistically from rocks that weathering had exposed in the chalk downs surrounding the site, or they may have been quarried deliberately from the Marlborough Downs some twenty-five miles north. What is in no doubt is the difficulty of transporting and erecting such gigantic mega-liths. Each weighs around twenty-five tonnes, is 2.1 metres wide and, at around 4.1 metres high, towers over a human.

Around thirty of the sarsons remain standing today. Most of them mark out an incomplete circle some thirty-three metres in diameter. This would be impressive enough, but the *pièce de résistance* is the stone lintels that sit on top of them. Today, six lintels remain in position in the outer circle.

There is an inner horseshoe of five trilithons, the name given to each pair of upright stones and its supported lintel. Three of these are still standing, although one was re-erected after the lintel fell in the eighteenth century. One of the other fallen trilithons has come to rest on top of the central 'altar stone'.

Among the giant sarson stones we find the bluestones of the original circle. Although most are fallen or damaged, they have been moved to mirror the new design, forming a con-centric circle between the sarsons and the trilithons, and an inner horseshoe between the trilithons and the altar stone.

Around the circle are five other sarson stones. Four mark the corners of a rectangle that encloses the stone circles, and are called station stones. The fifth is the Heel Stone, which lies to the north-east of the circle, the direction into which the trilithon horseshoe opens. The Heel Stone is what particularly draws attention to Stonehenge because when standing inside the circle at dawn on Midsummer's Day, the Sun rises close to – but not over – the Heel Stone. This misalignment may be explained by the fact that there was once a companion stone next to it, and the midsummer Sun rose in the opening – the gate – between the two stones.

Midsummer is the longest day of the year. In the northern hemisphere, it occurs around 21 June, and is also known as the summer solstice. The length of each day varies across the year depending on how the Earth's position in its orbit combines with the tilt on its axis. No matter where the Earth is in its orbit, the axis always points in the same direction. During the summer months in the northern hemisphere, the northern axis is tilted towards the Sun and the days get longer as the Sun climbs higher in the sky. Six months later when the Earth is on the opposite side of its orbit, the northern axis now points away from the Sun. This is winter time for the north, the Sun never rises very high and the days get shorter.

As the seasons change, so do the Sun's rising and setting points along the horizon. In the summer, the Sun rises in the north-east and sets in the north-west, which allows it to follow a longer, higher path across the sky. In the winter,

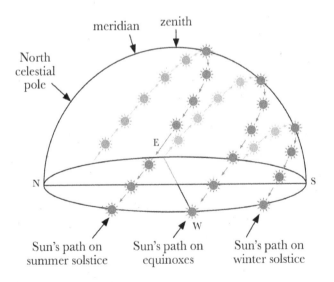

During the summer months, the Sun follows a longer, higher path through the sky than in the winter months.

the rising and setting points move to the south-east and south-west respectively and it never climbs as high.

The name solstice comes from Latin and means the point at which the Sun becomes stationary in the sky. On that day, the Sun will rise to the highest it is ever going to reach in the sky and the day will be the longest. After that, the Sun will not rise so high and six months later, the shortest day will occur. In the north, this midwinter point – the winter solstice – occurs around 21 December. The Sun reaches its lowest noon altitude before starting to creep higher into the sky again, each day.

The alignment of Stonehenge with the summer solstice was first brought to 'modern' attention in 1720 by the English antiquarian William Stukeley. Working at the beginning of the scientific revolution (he was an early biographer of Isaac Newton), Stukeley sparked interest in the idea that the huge monument was not just a ceremonial place but an astronomical observatory. Interpreting Stonehenge in this way culminated in the 1960s when the British-born astronomer Gerald Hawkins used a recently invented IBM mainframe computer to check all the various sight lines at Stonehenge for other celestial alignments. He programmed the machine with a model of Stonehenge, and the movements of the Sun and the Moon across the year. As the computer ran its program, it returned dozens of correlations. In contemplating these, Hawkins felt that the original fifty-six chalk pits could have been used as a means of predicting the lunar eclipses by moving markers from hole to hole. His findings were first published in the prestigious journal *Nature* in 1963,[16] and then more fully in a book called *Stonehenge Decoded* in 1965.[17] Almost certainly inspired by the fact that he had used a computer to find these alignments, he proposed that Stonehenge itself could be thought of as a Neolithic astronomical computer. His conclusion was backed up by other astronomers, including the high-profile polemicist Sir Fred Hoyle.

Criticism from archaeologists was sharp and immediate. They argued that the sheer number of sight-lines checked by Hawkins would be bound to reveal alignments just by chance. Over the centuries, bona fide academics and enthusiastic amateurs alike have proposed all sorts of possible interpretations of Stonehenge. In the Middle Ages, it was thought that the legendary wizard Merlin used it for magical purposes, and that a giant had built it. In more modern times, it has been variously proposed that the stones were erected to create a sonic 'laboratory', to recreate the shape of the female genitals as a fertility symbol, and (perhaps inevitably) as a landing pad for UFOs. In 1967, at the height of the debate about whether Stonehenge was an ancient astronomical observatory, British archaeologist Jacquetta Hawkes finally had enough. She wrote dismissively, 'Every age gets the Stonehenge it deserves – or desires.'[18]

These days, to avoid imposing their own wishes or modern ways of thinking on their interpretations, where there is a lack of written evidence, archaeologists look at the surrounding landscapes and structures for evidence to support their hypothesis. And in the case of Stonehenge, this provides some fascinating insights.

The neighbouring area is now known as the Stonehenge World Heritage Site. It encompasses 26.6 km² and contains more than seven hundred archaeological sites and monuments, more than half of which are burial mounds. So, clearly the site is an important funerary area. Yet there can also be no doubt that Stonehenge was deliberately aligned with the solstices.

In 2013, Professor Mike Parker Pearson of the Stonehenge Riverside Project announced the results of an excavation into the so-called Stonehenge Avenue.[19] This landform comprises a pair of parallel banks and ditches about twelve metres apart. They run for 1.5 kilometres past the Heel Stone in the direction of the midsummer sunrise. Parker Pearson found that underlying the avenue were natural ridges that had been cut into the landscape by meltwater at the end of the last Ice Age, about 9700 BCE. The coincidence that these water features lined up with the direction of the midsummer sunrise could be the reason that prehistoric humans developed Stonehenge as a sacred site.

Whatever the reason for it being sited where it is, it seems safe to say that this extraordinary area was a place of ritual and veneration of the dead, which involved celestial objects in some way or another. As with Nabta Playa, it feels religious. Yet, without written records, there will always be room for uncertainty.

★　★　★

For the first writing, we have to look east. Some time between 5500 BCE and 4000 BCE, the first cities were born in the so-called fertile crescent. This curving stretch of land begins at the Persian Gulf and runs along the Euphrates and

Tigris rivers to the Mediterranean Sea. It then follows the coast west, along the Levant, until it joins Africa, and from there it traces the Nile.

These river valleys were fertile because their annual floods left behind nutrient-rich soil that naturally grew crops. In the process of tending these fecund resources people settled permanently. As they became ever better at looking after the crops and encouraging them to grow, so the people developed agriculture, which gave them a certain mastery over the land. The permanence of their settlements allowed them to develop complicated social structures that we now recognise as the first instances of urban civilisation.

There were several places within this fertile crescent that developed independently of one another. Of these, one of the most significant was Sumer. Living between the Euphrates and the Tigris in Mesopotamia (today's southern Iraq), the Sumerians were the first to develop writing. In the triangular symbols of their cuneiform script we find the first religious texts, which are inextricably linked to the night sky.

The Kesh Temple Hymn was written on clay tablets by Sumerian scribes as early as 2600 BCE. There is no surviving whole tablet that contains all 134 lines of the hymn, and so our modern translations are the result of fragments being pieced together from different tablets from other Sumerian sites. These tablets have been dated to vary in age by eight centuries. The remarkable consistency in the overlapping portions of the text shows that it was an important work, copied almost verbatim for the better part of a millennium.

First, the hymn lavishes praise on the city of Kesh and its tall temple. It likens it to the Moon in the sky: a brilliant beacon against a dark backdrop. It implies that the temple is an earthly connection reaching up into heaven and down into the underworld, and states that Enlil, the chief deity of Sumer religion, appointed the temple as divine during the making of the world. Next, it boasts about the temple's ownership of livestock – a sign of great wealth – and finally fills in more backstory about the creation of the world and how great the temple is. Line 125 even asks: 'Will anyone else bring forth something as great as Kesh?'

Repeated without question as the hymn was endlessly copied, this story celebrated Kesh as a special place of chosen people. And the temple itself was identified as the most special place from which to contemplate the night sky and its meaning. In this we can see the kind of self-aggrandisement that characterised the secret societies of the hunter-gatherers writ large: the city itself is said to be a chosen place for its connection to the heavens, i.e. the night sky.[20] And it worked. The city gained great status and attracted inhabitants from far and wide. The people who administered the temple grew rich and powerful because their establishment owned vast swathes of land on which crops grew. It granted loans, employed citizens and maintained itself as a kind of clearing house for mercantile transactions.

Most significantly, a creation myth associated with the night sky is completely interwoven with the hymn's blatant self-promotion. The surrounding universe is clearly marked out as a divine realm, associated with Earth but utterly different from it. In his creation of our world, Enlil was said to have separated it from heaven. In so doing, he made Earth habitable for humans, and claimed it as his own. The goddess An took possession of the sky, but Enlil alone remained the connection or 'mooring rope' between the two realms because he is the air, out of which the stars and planets between heaven and Earth are made.

Clearly the religions of the third millennium BCE were extremely well developed. They are full of complexity and subtlety and were widespread, which indicated that they were not new inventions. These faiths were evolving entities that the written records captured in flight. We know this because in the centuries and millennia that followed, empires such as Assyria, Akkadia and Babylonia all rose and fell in the region. Each adapted the religion of the other. For example, in Babylon, the powers of Enlil became associated with the god Marduk.

Yet these various changes are mainly cosmetic, and the bedrock of Mesopotamian belief remains unchanged: the gods made order out of chaos by creating the world and everything in it, including us, and then embodying themselves within nature. And this is crucial because it means that when the wind blows, or an eclipse happens, it does so because there is divine will behind it.

From this point of view, the Sumerian religion – and those of the other Mesopotamian cultures that followed from it – are all early attempts at what we could term 'science'. Today we often portray science and religion as diametrically opposed, yet here in Sumer, the religion sprang from a desire to understand why things happened in the natural world. It was a rational thought process balanced on the assumption that supernatural beings created the universe. And by studying that universe, one could gain insight into the gods.

As the stars and the seasons suggested, Heaven and Earth were 'mirror-realms', one populated by gods and the other by humans – alike yet dissimilar. And the two were separated by a gulf that was impossible for humans to cross. But in the burgeoning civilisation of Ancient Egypt around this time, ideas began to change.

The Ancient Egyptian civilisation took firm root in the Nile valley around 3100 BCE. Their first writings date to this time and include religious texts that also link the night sky with the home of gods. Their earliest description of the night sky is as an enormous falcon stretched across it. The eyes of the bird were the Sun and the Moon, the white flecks on its breast and belly were the stars, and the winds of earth were created by the flapping of its wings. The god Horus, who is one of the Egyptian deities associated with the sky, was often depicted as a falcon-headed man.

The Egyptians also devised reasons for some celestial events, such as day and night. Ra, the sun-god, was said to travel in a celestial barge across the sky by day. He then transferred to a second barge and moved unseen through the underworld by night, returning to the opposite side of the sky in time for the next morning and another voyage in his celestial barge.

The Egyptians had such a wealth of religious stories associated with the night sky that sometimes they contradicted each other. For example, one of Ancient Egypt's most important sky deities was Nut. As goddess of the night sky, she was sometimes portrayed as a sacred cow, sometimes as a naked woman. In both incarnations, she stretches across the sky with her head in the west. No longer white flecks on a bird, the stars appear along Nut's belly, and Ra's barge traverses her torso during the day. According to this interpretation, Nut swallows Ra at sunset. He then travels back through the length of her body during the night and she gives birth to him in the east at sunrise.

While all of this is essentially a repackaging of Sumerian ideas, the Egyptians do make one significant change that echoes down the ages to today. It is the belief that the night sky is the final resting place for the souls of dead pharaohs.

In Sumer, the night sky was believed to be the exclusive realm of the gods, and Earth was the abode of humans. Even the King was merely a mortal. According to Egyptian thought, however, pharaohs were human incarnations of gods and so could ascend to the starry realm after death. This established a link between mortal humans and an after-life among the stars, which was a precursor to the concept of a celestial paradise or 'heaven' for everyone.

Belief in an afterlife was deeply embedded in the Egyptian psyche. They concocted an extraordinary set of beliefs and stories, metaphors and rituals around it. At first, these were aimed at the everlasting salvation of the pharaohs, who were buried in increasingly elaborate pyramids.

The best known of these are the three giant pyramids of Giza. Construction here began around 2550 BCE, when Khufu selected the site for his final resting place. It took twenty years for his pyramid to be built. With each of its four sides stretching 230 metres at the base, and topping 146 metres in height, the pyramid of Khufu is the largest of the three necropolises and remained the tallest man-made construction in the world for nearly four millennia. It was eventually beaten in 1092 by England's Lincoln Cathedral, which reached 160 metres.

Volumes have been written about the pyramids and their supposed alignments to the night sky. Of these various ideas, the only one that is incontrovertibly true is that the square bases of these three great pyramids are accurately aligned north–south and east–west. The famous Sphinx found at the site also faces directly east. This means that the Sun rises above its head on the spring equinox, around 21 March every year. How the Egyptians orientated these constructions so precisely has long been a matter of debate.

Today, if we wanted to build something pointing north–south and had no compass to hand, we would line it up with Polaris, the pole star. This fairly bright star sits very close to the projection of the Earth's north pole onto the sky. As the night progresses, and the Earth turns on its axis, Polaris appears to stay still while all the other stars pivot around it.

Back in Ancient Egypt of the third millennium BCE, however, the north celestial pole was far from Polaris. This is because the Earth wobbles like a very slowly spinning top in a phenomenon called precession. It takes 25,772 years for the Earth's rotation axis to complete a full circle. Around 3000 BCE, a dim star named Thuban, in the constellation Draco, which is just one-fifth the brightness of Polaris, sat closest to the pole. Some have suggested that the Egyptians took their sight lines from this, which seems a distinct possibility in their darker, unpolluted skies.

In 2000, however, archaeologist Kate Spence of Cambridge University, in the UK, used a computer to look

at the orientation of the night sky that the pyramid builders would have seen above them. A pair of reasonably bright
stars leaped out at her: Mizar in Ursa Major, and Kochab in
Ursa Minor. These two stars sat at almost identical distances
from the north celestial pole, and as the night went on, they
circled it like two cats sizing each other up for a fight. To the
Ancient Egyptians an imagined line between these two stars
would always pass through the north pole.[21]

To transform this into the astronomical equivalent of a
plumb-line, all they had to do was wait until the stars were
in a vertical line upwards from the horizon. This would have
happened at one point on every single night because of the
rotation of the Earth. Thus Spence showed that astronomers
could have performed a nightly check on the sight lines of the
pyramids. The hypothesis was widely reported by the media,
and much debated among Egyptologists and other academics.

As ever, though, since the pyramid builders left no written record of their construction and alignment methods,
we can't say for certain whether there is a true link between
the pyramids and the night sky. When we look inside some
other pyramids, however, the story is dramatically different.

Less than twenty kilometres south of Giza is Saqqara,
another pyramid complex. Although these are not as tall as
those in Giza, they are notable for other reasons. In 1881,
the French Egyptologist Gaston Maspero was investigating
the site. Something drew him to the ruined pyramid of Unas,
and while others had only examined the outside, Maspero
found the way in. He wound his way through its dark cor-
ridors until he found the burial chamber. A beautiful black
sarcophagus made of polished basalt lay beneath a vaulted

ceiling inscribed with stars. But it was the walls, covered floor to ceiling in hieroglyphs, which truly drew his attention.

When the translations were complete, the archaeologists read of the perilous journey that Unas's soul would have to make through the underworld before it could ascend to the celestial realm and join the procession of Ra in the afterlife. This was the equivalent of an instruction manual for the recently deceased pharaoh. His eternal soul would rise, read the instructions, and then set off on his journey to immortality.

Subsequent explorations of other pyramids found more examples of these 'pyramid texts', but Unas's tomb is the oldest known, dating to around 2320 BCE. They talk of the dead king's final destiny being to join the *Ihemu-seku* (imperishable stars) in the north of the sky. This is usually thought to mean the stars closest to the north celestial pole that are visible all year round, the so-called circumpolar stars.

While the king would end up in the stars, the fate of ordinary citizens was much less glamorous. They did not ascend after death but instead spent the rest of eternity either in a shadowy underworld or a mystical, fertile land of plenty called the Field of Reeds, where they continued a facsimile of their earthly lives. But this divergence in the fates of pharaohs and common folk would begin to close over the course of the Ancient Egyptian period.

The great age of pyramids and stability in Ancient Egypt, known as the Old Kingdom, came to an end some time around 2180 BCE. In the run-up to the collapse, the regional governors had begun to amass money and, with it, power. Termed nomarchs, they turned their administrations into hereditary positions, which allowed their subsequent families to grow further in influence and stature. And they began to exert more influence on the pharaoh himself, who at the time was Pepi II.

Evidently, Pepi II was not strong enough to corral his nomarchs, who fell into disagreements and war with each other. This chaotic period lasted for a century and a quarter, by the end of which two rival dynasties had emerged that fought for control of all Egypt. A leader called Mentuhotep II was victorious and became the first Pharaoh of the Middle Kingdom, but the turmoil had a profound effect on the ordinary Egyptians' view of pharaohs, who were no longer seen as divine and infallible. As evidenced by all the fighting, the position was clearly up for grabs. And that meant all the privileges of being a pharaoh were similarly within reach – including a place in heaven.

Around 2100 BCE, star tables were being inscribed into wooden coffin lids. By 2000 BCE, around fifty years after Mentuhotep II reunified Egypt, writings following a similar pattern to the pyramid texts began to appear in more ordinary grave sites. The coffin texts, as they have been named, are clearly designed to guide commoners to their celestial destiny with the gods. They developed into the widely known Book of the Dead. This was a set of funerary passages used widely during the New Kingdom of Egypt, beginning around 1550 BCE. In the Book of the Dead, the final destination of the soul is set out in a series of spells to which a newly deceased person must adhere. But there's a catch – not everyone will be admitted into the celestial realm.

Spell 125 recounts the ceremony of 'weighing the heart'. This is the first description in history of a supposed judgement by the gods following death to decide whether or not an individual's soul has earned a place in heaven. The Egyptian underworld is called Duat, and the recently deceased are taken there by Anubis, the god of mummification and the afterlife. In the presence of Osiris, the god of the underworld, a dead person must attest to having lived a life free from proscribed sins that range from the heinous to the hardly avoidable. The list includes: robbery, murder, witchcraft, adultery, debauchery, slander, cursing, eavesdropping and raising one's voice.

To test the accuracy of this statement, the deceased's heart is weighed by the gods on a set of scales against an ostrich feather, which is an embodiment of the goddess Maat, who stands for truth, balance, law and morality. Should the heart be weighed down by sin and so heavier than the feather, a fearsome creature called the Ammit – part lion, part hippopotamus, part crocodile – would eat the heart, dooming the deceased to restlessly wander the underworld for eternity.

If the heart weighed no more than the feather, then the deceased could join Ra as part of the crew of his celestial barge. In versions of the myth where the worthy souls took their place in the Field of Reeds, this paradisiacal afterlife was becoming ever more closely associated with the night sky. It became part of the story that inhabitants of the Field of Reeds could be called into the celestial realms to defend Ra against the forces of darkness. In this way, the celestial afterlife was democratised even though not everyone could get in.[22] This story also represents the invention of 'big' or moralising gods – the ones that sit in judgement of us upon our death and decide whether we take our place in heaven. And there is an even more explicit link to the night sky within this story than the simple concept of heaven.

Maat was also the goddess of the stars and the seasons on the Earth. She personified the actions of the deities who presided over creation. Whether by coincidence or design we will never know, but the religious linking of the fate of the dead to the seasons seems to mirror the practice at Stonehenge of simultaneously venerating the dead and charting the passing of one season to another. Was this how those ancient people thought of death: the passing of a season – one state transforming into another? It certainly seems possible and would give another reason to continue searching the night sky to add meaning to our earthly lives.

According to psychologist Ara Norenzayan of the University of British Columbia, Canada, belief in a judgemental god does not spring up by chance. Instead, it comes about as a means of fostering co-operation in a growing civilisation that needs to work together for the common good. The motivation for co-operation is clear: stick to the rules in this life, and be rewarded in the next.[23]

But what sparked civilisation in the first place? According to traditional thinking, agriculture was the catalyst. No longer would groups of humans wander the land, hunting wild beasts along the way or herding cattle from one pasture to another. Instead, they would settle and cultivate crops. In Egypt this move was driven by the annual flooding of the Nile, which brought fertile soil to the riverbanks. Eventually, increasing numbers of people lived together, forming the first cities, and with this change came the birth of judge-mental gods such as Maat to keep everyone in harmony.

The switch to agriculture is known as the neolithic rev-olution, a term coined by the Australian archaeologist Vere Gordon Childe in 1924.[24] In reality, it happened gradually over many centuries rather than swiftly as the word revolution might imply. Many reasons for the rise of farming have been suggested by academics over the years, their focus often being on natural climate change. Childe himself favoured a hypothesis that the land became drier and less hospitable, and thus forced people to live together around oases and rivers. A related idea links the mass extinction of wildlife, caused by the arrival of a warmer climate, to the switch from hunting to crop tending. In all of these cases, humans are being driven by forces beyond their control to adapt their survival strategies or die. But one recently discovered archaeological site in Turkey sug-gests something very different indeed.

★　★　★

Göbekli Tepe in south-eastern Turkey was first noted in a survey of the area during the 1960s but it was the mid-1990s before excavation began. Under the leadership of Klaus Schmidt of the German Archaeological Institute, scholars began revealing this extraordinary site. Dating to 9500–9000 BCE, it is the oldest example of monumental architecture in the world, built some eight thousand years after the cave paintings at Lascaux, and five thousand years before the first cities were founded in the fertile crescent.

Göbekli Tepe consists of at least twenty circular structures, the largest of which are around thirty metres across. Each is bounded by huge T-shaped stone pillars, around two metres in height. Only a few of the stone circles have been uncovered so far and, unlike the roughly hewn sarsons of Stonehenge, these pillars are exquisitely carved with animal depictions, which are often fierce and angry portrayals. In the centre of each circle, two larger megaliths stand to a height of four metres, presenting a sightline to the sky.

In what is now common practice, Giulio Magli, from the Polytechnic University of Milan, investigated what the night sky over Göbekli Tepe would have looked like to the builders of the temple. One thing leapt out at him.

Sirius is the brightest star in the night sky. It is outshone only by the Sun, the Moon, and the planets Venus and Jupiter. Even in today's light-polluted skies it is an unmistakable beacon. Yet from prehistoric Turkey, it would not always have been visible. The 25,772-year cycle of precession causes the direction of north to change in the sky, and this makes stars rise and set at different times. Some stars will even disappear from view altogether during parts of the cycle, only to appear again centuries or millennia later. Sirius was one such star, and what caught Magli's attention was that it returned to visibility in the south-eastern skies of Turkey around 9300 BCE, right in the middle of the date range for the founding of the temple.

He suggests that the return of Sirius to the sky captured the people's imaginations so completely that they erected the temple to follow the 'birth' of this star into the night sky. Furthermore, he thinks tracking Sirius was the reason why there was more than one circle built here. As the centuries passed and Sirius rose from different positions along the horizon, so those who came to the location would periodically build a new circle to keep track of the star's progress. Of the currently excavated rings, three appear to align with Sirius's rising position for 9100 BCE, 8750 BCE and 8300 BCE.[25]

Today the landscape is a barren desert, but back in the time of its founding the area would have been lush with plants, fields of wild wheat and grazing animals. On the face of it, a prime place to settle. But the archaeological evidence tells a different story: the stone circles are the earliest constructions at this location. There was no settlement beforehand, no nascent city that then constructed a convenient place to worship. And the clinching evidence for this interpretation can be found in the animal bones that litter the site.

The Belgian archaeologist Joris Peters has studied more than 100,000 bone fragments from Göbekli Tepe. Many shows signs of being hunted, butchered and cooked, but the crucial observation is that the animals are all wild species. According to Peters, this appears to suggest that hunter-gatherers built this place for the sole purpose of religious worship rather than settlement.[26]

Göbekli Tepe may originally have been the site of an annual gathering between nomadic groups, which grew in importance once Sirius had been spotted above the horizon. Certainly the investment of time and effort in constructing each stone circle would have been considerable. Each stone pillar alone weighs up to ten tonnes. Although they were cut from nearby cliffs, hundreds of people would have been needed to transport and erect them into each stone circle. Needing that many people on hand over an extended period of time could have led to semi-permanent encampments that eventually turned into the settlements that archaeologists now find surrounding the site.

This upends the idea that city living and civilisation brought about that spare time needed to build structures unrelated to simple survival: i.e. ceremonial places. Instead, it suggests that here the construction of a ritual site possibly dedicated to the night sky was one of the catalysts for urban settlement because it drew people to the area. Indeed, before his death in 2014, Schmidt came to the conclusion that Göbekli Tepe was a place of refuge. If true, the conclusion is staggering: our primal fascination with the night sky and the hope that it might answer our questions about existence helped lead to civilisation itself.

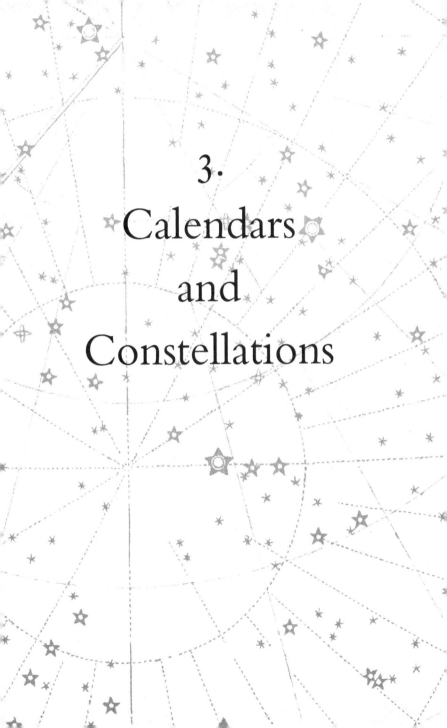

3.
Calendars
and
Constellations

While we cannot be sure whether agriculture was a cause or an effect of settled living, there can be no doubt that tending crops changed us for ever. French anthropologist Jean-Pierre Bocquet-Appel studied human remains in 133 ancient cemeteries that straddled the neolithic revolution and concluded that there was a population explosion at this time.[27] It was fed (literally) by the greater amounts of food that could be accessed by domesticating crops. Instead of living as hunter-gatherers wandering the landscape to see what chance presented, humans started to take some control of their surroundings.

The one factor that they couldn't control, however, was the weather. Agriculture came with an inherent risk: inclement seasons could ruin whole crops. Unlike their hunter-gatherer forebears, the early agriculturalists were tied to a single spot, and each year they would gamble on whether they could protect and urge their crops to fruition. As a result, agriculture redefined our relationship with nature, and helped drive a deeper appraisal of the night sky and its supposed relationship with Earth.

As we have seen, the various henges and other monuments of prehistory all served a dual purpose as places both of veneration and sky watching. Following the neolithic revolution, veneration grew into worship and then organised religion; likewise the observance of the solstices developed into full-blown timekeeping with the inception of calendars. But these two endeavours were not separate. In keeping with the ideas of animism – that the heavens and the Earth were joined in a spiritual way – calendars were originally developed for religious ends.

The first formal calendar found in the archaeological record comes from Sumer in the second millennium BCE. It is known as the Umma calendar of Shulgi, a Sumerian king who lived in Ur, a city-state located in modern-day Iraq. As described in the previous chapter, the Sumerians believed that the gods fashioned the earth and the surrounding universe by making order out of chaos. According to the tale, the gods then created humans by mixing the blood of a slain god with clay. They did this so that the humans could perform the work on Earth that the gods had grown weary of doing themselves. Thus, with the work entrusted into human hands, the gods then lived lives of leisure by embodying themselves in the various celestial objects and the forces of nature.

With this as the Sumerian core belief, the observation of nature became nothing less than the contemplation of divine order. To chart the seasons of Earth, the changing of the constellations through the year, and the various comings and goings of the other celestial objects, was to know the will of the gods. And the regularity of these comings and goings meant that in the creation of a calendar, the Sumerians were capturing the grand design. It was the equivalent of a holy book.

The Sumerians based their calendar on the Moon. A new month was announced after the first sighting of the slender crescent of a new moon on the western horizon just after sunset. King Shulgi built the great ziggurat at Ur so that priests could look for this event from the top of its elevated platforms. In addition, this large stone building with its sheer walls and extended stairways served as a seat of power for his administration, and a shrine to the Sumerian Moon goddess Nanna.

By the time of the Neo-Babylonian Empire's rise to power in the seventh century BCE, the origin of the calendar had been enshrined in a creation myth known as the Enuma Elish. Borrowing heavily from an earlier Sumerian version, it relates how the god Marduk created cosmic order and set himself up to rule heaven, leaving the Babylonian emperor to be his deputy on Earth. In the course of setting up this cosmic order, Marduk was said to have placed the Moon in the sky to measure out the passing of time by its phases.

Indeed, the Moon is an excellent starting point for sub-dividing the year because of its cycle of phases. Over the course of 29.53 days, it grows from nothing to full illumination and back again, before beginning the cycle once more. The lunar phases are a result of the Moon's orbit around the Earth and the fact that it gives out no light of its own, but simply reflects sunlight. As the Moon moves around the Earth, the angle it makes to the Sun changes from our perspective and we see larger areas of illumination. At full moon, it is lying in the opposite direction to the Sun, and so we see a fully illuminated hemisphere. A new moon takes place when the moon lies in the same direction as the Sun. In this position, the Moon is blocked from our view by the

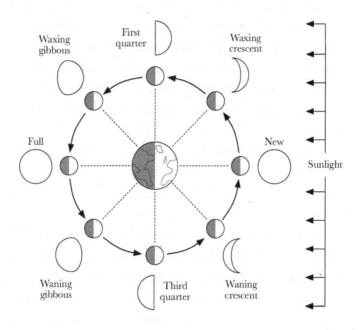

The lunar phases are caused by where the Moon happens to be in its month-long orbit of the Earth.

Sun's glare. Even if we could see it, we would just be looking at a dark face, as the hemisphere pointed towards us would be in shadow.

The month represents a single day and night cycle on the Moon. Whereas the Earth spins on its axis once every twenty-four hours, the Moon is locked into rotating on its axis only once during its orbit around our planet. It wasn't always like this, but the Earth's gravitational grip that holds the Moon in orbit has also sapped it of its ability to spin any faster. Only being able to spin once during its orbit explains why the Moon has become locked into always showing the same face to Earth. Thus the lunar phases represent a day and night cycle on the Moon. From dawn to noon on the Moon takes seven Earth days. Then a similar time to sunset, followed by a fortnight-long night.

While lunar calendars are great starting points, they are not ideal. The issue is that twelve lunar months (lunations) only equal 354.36 days and so fall short of a full year by about ten days. This means that a year defined as twelve lunations would fall out of step with the seasons because these are defined by the position of the Earth in relation to the Sun, and it takes the Earth a full 365.25 days to complete one orbit. For this reason, purely lunar calendars gradually stopped being used to chart the year from around the sixteenth century but continued to be used for setting religious events. To this day, Islamic society uses a purely lunar calendar to set the dates of religious rituals including Ramadan, the month of fasting, and the Hajj, the annual pilgrimage to Mecca. As a result, the rituals always start ten days earlier each successive year.

To reconcile a lunar calendar with the year, it is necessary to employ a system of leap-months. This solution is known as the lunisolar calendar. In this system, the disadvantage is that the years have a variable length. Typically there are two years lasting twelve lunar months followed by one containing thirteen lunations, which makes up the shortfall and resynchronises the calendar with the seasons. The Jewish religious calendar in use today is a lunisolar calendar, as was the Ancient Mesopotamian one. After dabbling with a lunar calendar, the Ancient Egyptians went on to develop one based solely on the Sun, and solar calendars are still used today across most of the world. Even in Islamic countries, solar calendars are used for non-religious purposes.

★ ★ ★

In Ancient Egypt, the seasons were all based upon the annual flooding of the Nile. Every year between May and August, a tremendous monsoon would engulf the highlands of Ethiopia, which lies to the south of Egypt. This rainwater would feed Lake Tana, and from there it would funnel through the so-called Blue Nile into the River Nile proper. The inundation would cause the Egyptian Nile to overflow and flood the surrounding lowlands. It began with a steady rise in river level in June, which would suddenly increase in mid-July, covering the lowlands with the fertile soil that made crop rearing possible. This natural cycle of flooding – and its inherent uncertainties – was only broken in the 1970s with the completion of the Aswan High Dam. With little change in the arid weather conditions throughout the year, the Egyptians made use of this cycle to divide their year into three seasons. The year began with Akhet, the season of the inundation; Peret was the season of growth; and Shemu was the harvest time.

By coincidence, there was a predictable celestial event that corresponded with the annual flooding and the Egyptians used this as the marker of their new year. The largely cloudless skies of the Nile valley meant that the people who settled there were well aware of the solar, lunar and stellar motions above them. And the event that caught their attention involved Sirius – or, as the Egyptians called it, Sopdet.

58

Earth's tilted rotation axis means that, as in south-east Turkey, Sirius is not visible all year round from Egypt. It spends seventy days of the year below the horizon, and then by pure chance returns to visibility in the run-up to the flooding of the Nile. It seemed too much of a coincidence for them to ignore, and as if to confirm the apparent link between earthly and celestial events, Sirius is located in the sky near the misty band of the Milky Way, which the Egyptians equated to the Nile.

Because its appearance heralded the flood, Sirius became associated with Isis, the Egyptian goddess of rebirth and maternal love. Her tears for her slain husband, Osiris, were said to have filled the Milky Way, and the annual inundation of the Nile was thought to be the earthly manifestation of the return of her tears. Sirius was also associated with the Egyptian goddess Hathor, who symbolised fertility, and this gave rise to a different myth about the inundation. In this myth, Hathor, symbolised as a cow with Sirius shining between her horns, is sent by Ra to punish humankind for a planned insurrection against his rule. Enraged, Hathor begins to massacre humans indiscriminately, so Ra floods the land with red-coloured beer to prevent her from killing everyone. In her rampage, Hathor mistakes the beer for blood and drinks deeply, becoming soporific and pacified. Because the Nile carried sediments that turned it red during the flood, it was associated with the return of the goddess.

In 1890, British astronomer Norman Lockyer set out to explore Egypt. He was intrigued by the reports of celestial alignments and after research of the area proposed that the Temple of Isis at Dendera had been built to look for

Sirius's return to the night sky.[28] Another temple to Isis is located on the island of Philae, in the Nile. Near to it is a temple to Satet (a goddess also associated with the flood myth), which has been suggested as the very site where the Ancient Egyptians watched for the return of Sirius in order to declare the new year and prepare for the flood.[29] Near the appointed time of year, astronomer-priests would gather in the twilight hours of the morning to watch the eastern horizon, waiting for Sirius to appear just before dawn obscured the view. On the morning that they saw it, so they declared that a new year had begun. This moment was called the heliacal rising of Sirius.

Each of the Egyptian seasons was divided into four months, each thirty days long. Each month was then divided into three ten-day periods known as decans. Because twelve thirty-day months only give 360 days to the year, an additional short month of five days was included in the Egyptian calendar to keep the seasons in step with the calendar. Every fourth year, this short month was extended to six days in the Egyptian equivalent of our leap year. But the Egyptians also maintained their earlier lunar calendar for religious purposes.

The seventy days in which Sirius stayed below the horizon at night were also thought to be religiously important. The Egyptians considered such a period to be the duration of a soul's journey through the underworld, and they therefore used it as the length of time allowable in which to embalm a body and mummify it for burial.

★ ★ ★

As well as religious events, astronomy was also used to define time-keeping by subdividing the day around three major astronomical moments: sunrise, midday, sunset. Our further subdivisions of the day into twenty-four hours and the hour into sixty minutes both derive from the early civilisations.

Sundials were in use by around 1500 BCE in various parts of the fertile crescent. In 2013, the base plate of an Egyptian sundial was found in the Valley of the Kings by researchers from the University of Basel, Switzerland.[30] Unearthed in an area of stone huts used by the workmen who constructed the tombs, the artefact is notable because the semicircle drawn on the base plate has been subdivided into twelve roughly equal segments to delineate twelve 'hours' of daylight. The question is, why twelve?

The use of a base twelve (duodecimal) counting system was established in Ancient Egypt, and it is generally thought that this was because twelve was the number of complete lunar cycles in a year. If nature had chosen to divide a year into twelve, then it seemed like an elegant piece of symmetry to do the same for a day.

At night, with no sun to guide them, the Ancient Egyptians defined a number of recognisable patterns of stars or single bright stars they would watch for on the eastern horizon. As each pattern, known as a decan, rose, the astronomers could tick off another 'hour' of the night. These decans can be thought of as early constellations. There were thirty-six in total, about a third of which would be used in any given night to give twelve roughly equal hours, no matter what the season.[31] And it is from this ancient root that we find the basis of dividing the day into twenty-four hours.[32]

Intriguingly these decans form the star tables that were painted on the insides of wooden coffin lids from around 2100 BCE. Perhaps they were designed to help the dead navigate their way to paradise. Much later, in the second century CE, a description of the use of decans is found on papyri in a document called 'The Fundamentals of the Course of the Stars', which is now usually referred to as *The Book of Nut*, after the name of the sky goddess. However, the lists of decans found in these various places differ from one to another, and are rarely associated with charts depicting their star patterns. This makes identifying them with the modern constellations difficult.[33] But in the archaeological remains of Dendera, about sixty kilometres north of Luxor, an amazing find was made.

The Temple of Hathor, Dendera, was discovered in the nineteenth century by Vivant Denon, a French archaeologist who accompanied Napoleon during the Egyptian campaigns. It was during this time that the extraordinary cultural treasures of Ancient Egypt were brought to Europe's attention. In the ceiling of the temple's portico was a bas-relief showing a detailed pictorial representation of the night sky, including the thirty-six decans. It would have been the first thing visitors saw if they looked up before entering the temple. Now it is something that visitors to the Louvre museum in Paris see, because although Denon initially contented himself with drawing the extraordinary sky map, after he became the founding director of the Louvre, he dispatched a stonemason to Egypt in 1820 with a sack load of tools and a modicum of gunpowder. These were used to remove the relief, before transporting it back to the French capital, where it has remained ever since.

Each decan is represented by an image of a god standing in a boat, surrounded by a small pattern of stars – although in a few confusing cases there are no stars surrounding the god. The boat is significant because this is the symbol used by the Egyptians to signify the movement of the stars across the night sky. Hence, the decans themselves are not thought to be full-blown constellations as we think of them today; instead each was a compact configuration of stars that would all rise at the same time. These asterisms appear to be composed of up to four stars each, yet unfortunately are not sufficiently defined to allow them to be associated with visible stars today.

But they do allow us to draw other important conclusions. First, they are yet another clear example of ancient humankind's practical and religious relationship with the night sky. Second, the Egyptians named the decans figuratively rather than descriptively, and this gives us a clue about the naming of the constellations that we use today.

It is often written that the constellations were named according to what the pattern looked like. While it is not difficult to see a man in Orion, and a lion in Leo, it is all but impossible to see a goddess in Andromeda, or a set of scales in Libra. And while Sagittarius is supposed to look like an archer (often depicted as a centaur), in reality the star pattern looks more like a teapot – even astronomers call it by that name.

In the naming of the decans, however, we see that the figurative names the Egyptians gave to these star patterns were probably an *aide-mémoire* to remembering what was essentially a celestial co-ordinate and time-keeping system. And this brings us to the origin of the constellations.

*   *   *

At the Lascaux caves in France, the walls are covered in nearly six thousand figures. These can be grouped into animals, humans and symbols. Of the nine hundred or so animals, almost half are horses, but it is the bulls that draw our attention. In one section of the cave, four black aurochs (the now extinct ancestor of domestic cattle) dominate the walls. Of these, the biggest bull is 5.2 metres long. Above its shoulder is a pattern of six dots that will look eerily familiar to any northern skywatcher. They resemble a cluster of stars that can be seen easily with the naked eye. Known as the Pleiades, they are the only obvious tight-knit collection of stars in the whole sky. When they were painted on the cave walls they may also have been seen as heralds of spring and autumn. At that time, they would have been visible throughout the night during springtime, setting in the west just as the Sun was rising in the east. By the autumn, however, they would only have been briefly glimpsed rising at dawn.

But the most interesting part of the story is that today the Pleiades form part of the constellation we call Taurus, the Bull. They even sit just above the bull's shoulder, as they do in the Lascaux painting. So when we stand out under the stars today, look up at the Pleiades and imagine a celestial bull, we are going through the exact same thought process as that prehistoric artist some nineteen thousand years ago.

The Pleiades are also represented on a beautiful archaeological artefact called the Nebra sky disc. Found in Nebra, Germany, it has been dated to around 1600 BCE and consists of a bronze circle some thirty centimetres in diameter. Its blue-green patina is inlaid with gold symbols that seem to be clear representations of celestial objects. The most obvious of these is a full circle and a crescent depicting the Sun and Moon, and between them a grouping of stars that are generally considered to be the Pleiades. Though the exact astronomical purpose of the Nebra sky disc will almost certainly never be known, the fact that it contains a representation of the Pleiades is nonetheless a clear indication of the significance of that star cluster to ancient humans.

Given their singular nature, it is unsurprising that the Pleiades crop up in many folk tales associated with early people. What appears too striking to be pure coincidence, however, is that all these tales seem related. Aboriginal peoples from Australia, North America and Europe almost always identify the star cluster as a group of women. And often the associated story has them being chased by lusty men. One of their number succumbs, and disappears from the group. This explains a perennial mystery associated with this star cluster, namely that although only six stars can today be seen with the naked eye, it is often called the seven sisters.

The similarity of the stories suggests that the tale was concocted before our ancestors began their migration around the planet. And that means humans were projecting their imaginations into the night sky and telling stories about the stars many tens of thousands – or even a hundred thousand – years ago.

Similarly, the stars surrounding the Pleiades are usually associated with a bull. To the Ancient Sumerians it was the 'Bull of Heaven', from the Epic of Gilgamesh, one of the earliest pieces of great literature known. The epic poems developed over many centuries from individual folk tales of Gilgamesh. The bull crops up when our eponymous hero spurns the advances of the goddess Inanna. Hurt by his rejection of her, she sends the bull to kill him. It fails after an ally of Gilgamesh rips the bull in two and throws it back into the heavens. Even to this day, Taurus is usually depicted as the front half of a bull only. According to the Babylonians, the two hind legs could be found in the constellations we now call Ursa Major (the Plough) and Ursa Minor.

The Sumerians also recognised Gilgamesh in the stars. They associated him with the constellation we now call Orion, the hunter. They described a sword hanging from his belt, just as we also imagine for the modern incarnation. While it is true that the pattern of stars in most constellations do not look like their descriptions, Orion is one of the exceptions. In those bright stars, it is easy to imagine a belted figure holding up a shield or a bow to the neighbouring constellation of Taurus.

Stories like this have led to the idea that the constellations were devised by ancient people telling stories around a camp fire. According to this 'picture book' hypothesis, shepherds, minstrels or village elders whimsically looked into the sky and made up stories for the amusement and education of their communities.

While a few constellations such as Taurus and the Plough, and the constellations relating to Greek myths such as

Andromeda, Cassiopeia and Perseus, could have sprung from myth-making and story-telling, the decans show that grouping the stars into recognisable patterns means that they can serve a practical purpose by being a pictorial co-ordinate system. Nowhere is this more obvious than in the zodiac.

★ ★ ★

The zodiac is a band of constellations that circle the entire sky. The name is derived from a Greek term that means 'circle of little animals', although not every constellation is an animal. The constellations themselves are perhaps most familiar as our 'star signs' because they were given special meaning by the astrologers, as we will see in chapter 10. For now they are important because, like the decans, they can be used as a pictorial co-ordinate system. The twelve zodiacal constellations are Aries, Taurus, Gemini, Cancer, Leo, Virgo, Libra, Scorpius, Sagittarius, Capricornus, Aquarius and Pisces. They are defined by the path that the Sun, the Moon and the planets take across the sky. These celestial objects all follow more or less the same path; the only thing that varies is the rate at which they progress. Thus, the Sun, the Moon and the planets can be located in the sky by referring to which of the twelve zodiacal constellations they are currently traversing. For example, Mars in Taurus, Jupiter crossing from Sagittarius to Scorpius.

Although the twelve zodiacal constellations were not fully defined until around the fifth century BCE, the earliest known Babylonian star catalogue, which dates to the twelfth century BCE, already includes an early representation of them. In the MUL.APIN, a Babylonian star

This frontispiece to Epitoma in almagesti Ptolemei by Johannes Müller von Königsberg (1436-1476) depicts the author discussing astronomy with Ptolemy. Above them, the celestial sphere is marked with the zodiacal constellations. (AF Fotografie/Alamy)

catalogue from 1000 BCE, the zodiac is split into eighteen constellations, or 'stations' as they are called. The MUL. APIN is notable because it is a lot more than a simple list of stars. It is actually a manual for telling the time and charting the progress of the year from the night sky. It lists constellations that will rise and set simultaneously, pairs of constellations where one will rise when the other reaches the zenith, the paths of planets and the Moon through the zodiac. It also contains information about how to adjust the calendar to keep it in line with the rising and setting of certain stars, how to tell time from a shadow stick, and how the length of the night varies throughout the year.

It is also notable because it singles out four zodiacal constellations that mark four special points of the year. MULGU4. AN.NA, the steer of heaven, or as we recognise it today, Taurus; MULUR.GU.LA, the lion, our modern Leo; MULGIR. TAB, the scorpion (Scorpius); and MULSUḪUR.MAŠ, the goat fish (Capricornus). The points these constellations mark are the spring equinox, the summer solstice, the autumnal equinox, and the winter solstice – the major moments of the year marked at many of the megalithic monuments and caves of prehistory. By the time of the MUL.APIN, however, we see that the night sky is now used to chart the months, weeks, days and hours between these major moments. From those early Stone Age sightings, a sophisticated system of astronomical timekeeping developed with the zodiacal constellations at its heart, referred to figuratively as a way to distinguish them in these observations.

Two hundred years after the MUL.APIN, by the eighth century BCE, the Greek poets Homer and Hesiod were referring to the more recognisable constellations such as the Great Bear and Orion in their works. They also mentioned the bright stars Arcturus and Sirius and the Pleiades.

By the third century BCE, the Greek poet Aratus wrote a poem versifying astronomical knowledge. The *Phaenomena* is thought to rely heavily on the work of Eudoxus of Cnidus, a Greek astronomer who had studied the stars about a hundred years earlier. It is designed as an introduction to the night sky and its uses. It describes forty-seven constellations, most of which would be familiar to us today. Aratus explains how the sky revolves throughout the night and throughout the year, listing the constellations' various annual risings and settings. By this time, the eighteen stations of the Babylonian zodiac have been whittled down and combined into the twelve with which we are familiar today.

There are now eighty-eight recognised constellations, which were ratified in 1928 by the International Astronomical Union. Of these, more than half derive from a list devised in CE 150 by the Greek astronomer Claudius Ptolemy. He worked in the Roman province of Alexandria, Egypt, and in his great work, *Syntaxis* (more commonly known today by its Arabic name, *Almagest*), he catalogued one thousand stars and grouped them into forty-eight constellations. It was a major synthesis of astronomical knowledge and became the standard scientific reference work for the next millennium and a half. It enshrined the theory that Earth was the fixed centre of the cosmos, and that everything including the Sun revolved around it. This astronomical model remained in use until the sixteenth- and seventeenth-century work of

Copernicus, Galileo and others forced us to look at things differently, as we shall see in chapters 6 and 7.

Although Ptolemy's model of the universe didn't survive, most of his constellations did. Only one is notably absent from today's official list. It is *Argo Navis*, a giant constellation in the southern hemisphere that represented the Argo, the ship used by Jason and the Argonauts. In the burst of astronomical zeal that took hold in the scientific revolution of the seventeenth and eighteenth centuries, astronomers separated the ship into its component parts, and we now recognise Carina, the keel, Puppis, the poop deck, and Vela, the sails.

On the face of it, the evidence seems clear: the Greeks defined the system of constellations that we use today. But it is not quite so straightforward. Analysis shows that the description of the night sky by Eudoxus, which was enshrined in verse by Aratus, was one that neither man could possibly have seen. The southern constellations they talk about are simply not visible from Greece, and that means that the constellations we use today were not Greek inventions. Instead, they must have been received wisdom from a previous age and a previous people. But who?

To unpick this mystery, scholars have turned to the archaeological evidence, our knowledge of astronomy and some computer simulations about how the night sky would have looked in the past.

★　★　★

In Egypt, the Dendera sky map from the Temple of Hathor, which contained the bas-relief of the decan images, also showed the constellations. There are seventy-two individual pictograms that include Ptolemy's list of forty-eight, and the modern constellations of the zodiac are easily recognisable. Since Egypt lies to the south of Greece and has a more ancient civilisation, dating the sky map could be the key to the puzzle.

Indeed, when the French took possession of the sky map in the early nineteenth century, an intense debate about its age broke out. Many of the early estimates dated it to thousands of years BCE, making it a spectacular find indeed by proving that the constellations had been well defined much earlier than first thought.

However, the truth revealed itself through a study of the hieroglyphs that surrounded the zodiac. The crucial piece of evidence was the hieroglyphs found inside oval outlines called cartouches, which signified the name of the pharaoh at the time of construction. In the case of the Dendera zodiac, the cartouches were empty.[34] This placed the construction to a very specific moment in Egyptian history: the interregnum between the death of Cleopatra's father in 51 BCE and the joint accession of Cleopatra and her son with Julius Caesar in 42 BCE. In those nine years, any monument that was constructed had the name of the pharaoh left blank on its cartouches. Hence, the Dendera sky map comes about two centuries before Ptolemy and his *Almagest*, but a few centuries after Eudoxus and Aratus. While this particular monument can shed no light on who defined the constellations, another – more detailed – artefact has proved much more useful.

In the Museo Archeologico Nazionale, Naples, there stands the statue of the Farnese Atlas. It dates to the second century, around the time of Ptolemy's *Almagest*, and depicts the titan Atlas labouring under the weight of the sky, represented by a globe held on his shoulders. The statue is notable for being the oldest known celestial globe. Most of Ptolemy's constellations are depicted, but not as stars, rather as their pictograms, and these pictograms are reversed from how we imagine them looking up into the night sky. The reason is that we are seeing the night sky from 'outside', looking at it as a globe with the Earth as a dot imagined in its very centre. It is truly a god's-eye view of the sky. Crucially, where Atlas bears the globe on his shoulders, near to the point of the south celestial pole, the sky is blank; there are no constellations shown.

Similarly, in the poem of Aratus, there are no constellations listed near the south pole either. There are six constellations that form a ring around this blank area, and this is a major clue to the location of the people who defined the constellations.

Unless one is standing on the equator, there will always be a zone of stars that never set, and a zone of stars that never rise. In other words, anyone standing in the northern hemisphere will not be able to see all of the southern stars. So the gap around the south pole tells us instantly that the constellation makers lived in the northern hemisphere, and the size of the gap then tells us how far north.

If one were to stand at the north pole, one would only ever see half the sky. Stars would not rise and set; instead they would revolve around the north pole, pointing straight up into the night sky. So, if the original star map had only filled the northern half of the globe, then the astronomers in question would have lived in the Arctic. If they had lived on the equator, the globe would have been fully covered because they would have been able to see the whole sky throughout the year.

By analysing the size and shape of the uncharted southern sky region, a number of scholars have deduced that the constellations described in the poem originated in the minds of the Minoans of Crete during the third millennium BCE.[35] This is indeed a tempting conclusion, because the Minoans were a great seafaring nation who depended on the night sky for navigation.

Studies by Mary Blomberg and Göran Henriksson, both of Uppsala University in Sweden, have revealed that two Minoan structures – the palace of Knossos and the sanctuary on Petsophas – show striking celestial alignments in their architecture.[36] The palace allowed sunlight to illuminate its so-called Corridor of the House Tablets at dawn during the two equinoxes of the year. At the sanctuary, most of the walls are aligned with reference to the star Arcturus. From Crete in 1900 BCE, when the sanctuary was built, this bright orange star was not visible all year. One set of walls points to its heliacal rising position on the eastern horizon, and the other to its heliacal setting on the western. The importance of this star to the Minoans is that its visibility coincided with the good weather and thus the sailing season in the Mediterranean.

Away from land, the night sky is the only reference a ship's pilot would have apart from knowledge of sea currents and prevailing wind directions. Since the latter are subject to local disturbances, the stars are the fixed reference points, especially as there are few cloudy nights during the Mediterranean summer. But to steer a ship successfully demands an extraordinarily detailed knowledge of the various stars and the way they change their positions both throughout the night and over the course of the sailing season.

Even today, there are Polynesian navigators who are capable of steering their ships accurately across large stretches of open water using no instruments, only their knowledge of ocean currents, winds and the night sky. It is an extraordinary feat of delicate perception and judgement. In the UK and the United States, the *Nautical Almanac* is published every year and contains the positions of fifty-eight selected navigation stars that can be located with a sextant and whose altitudes can be used to calculate a ship's position at sea.

Without written tables, we can imagine the Minoans grouping stars into constellations that would help them remember sailing directions at various times of the season, rather as the Egyptians grouped stars into the decan patterns. Blomberg and Henriksson computed that Orion and Sirius would have marked the way from Crete to the Nile delta in September during the early second century BCE. Certainly by the time of Homer (around 700 BCE), this method was being described explicitly. In his epic poem, the *Odyssey*, he describes the hero Odysseus being given the advice to keep the stars of the bear (Ursa Major) on his left in order to sail eastwards.

Put this way, it all seems so plausible that the Minoans defined our constellations some time in the third millennium BCE. However, a more recent analysis by Bradley Schaefer, of the University of Texas, suggests a different location and a more recent origin.

He concentrated on the six most southerly constellations in the Greek depiction of the night sky, and used a computer

to calculate the most northerly latitude at which each constellation could be seen for each year back to 3000 BCE.
What was immediately noticeable in his work was that all
six constellations could have been seen in the sky from a
location at 30–34° north in around 900–330 BCE.[37]

This suggested someone at that latitude and at around
that time created the constellations. It could not have been
the Greeks, because the southernmost point of the country
is Cape Matapan at 36.4° north. And so these constellations
would never have risen above the horizon. Nor could it have
been the Minoans, as Crete is similarly located outside the
realm of visibility at 35° north. Also, their civilisation had
gone into decline after 1450 BCE following a succession
of natural disasters in the form of an earthquake and erup-
tions of the Mount Thera volcano. Instead, Schaefer points
to Babylon, which sits at 32.5° north and flourished until
around 540 BCE.

This doesn't rule out the Minoans as having had a hand in
defining our constellations. They could have begun the process of filling in those parts of the sky that prehistoric myth
had not touched. And then, as Schaefer concludes, the southernmost constellations were bolted on as late Babylonian
inventions. In this version of events, the night sky as presented
by Eudoxus and Aratus was a grand synthesis of ideas that had
evolved over millennia.

Regardless of the precise genesis of the constellations,
one thing is abundantly clear: the decans, the zodiacal constellations and the navigational uses of the stars all show
that, for the most part, the constellations were not a picture
book for idle amusement. They were defined as a celestial

co-ordinate system for the purposes of time-keeping and navigation. Yet, intertwined with these purely practical uses, the religious idea that the night sky held meaning for us, and that the gods wrote messages up there for us to interpret, continued to grow.

By classical Greek and Roman times, which stretched from the eighth century BCE to the sixth century CE, the practical and the religious aspects had begun to merge. Some of the important stars began to be seen not just as messengers but as causes of the events they heralded. For example, the Minoan association of Arcturus with sailing became transformed into a superstitious fear of stormy weather if the star was visible. As for the brightest star, Sirius, its glittering appearance in the summer months was thought to add to the power of the Sun, causing temperatures in the Mediterranean region to soar. Sirius is located in the constellation Canis Major, the great dog, and this is where we get our expression 'the dog days', meaning the hottest days of summer.

It's not difficult to see how they could arrive at such thoughts, given the common-sense observation that warmth and light come from the Sun, thus proving that a real, physical link exists between the heavens and the Earth. And as the ancient civilisations flourished, it became natural to try to reconcile these different aspects of the world, to look for something that could tie the two realms together and link us more directly to the night sky. The theory that sprang from this quest dominated human thought for almost two millennia, and continues to hold some in its thrall even today.

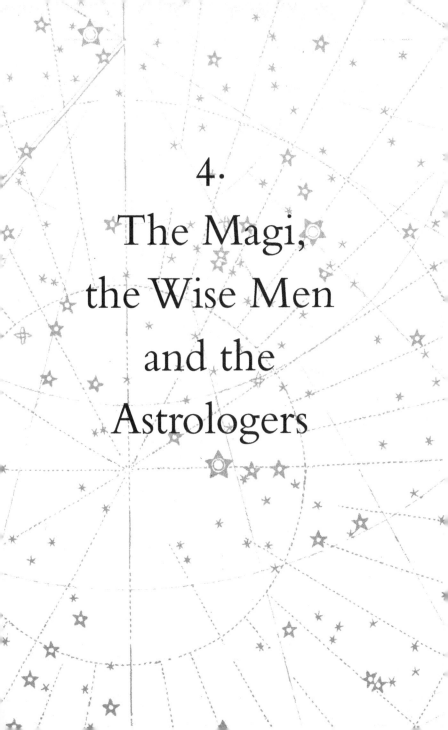

4.
The Magi, the Wise Men and the Astrologers

At the end of the MUL.APIN, the Babylonian star catalogue from 1000 BCE, there is a summary of how to apply astronomical knowledge to everyday life. It comes in the form of omens that can be read from the appearance of various stars, planets and meteorological phenomena such as sudden winds. These passages are significant because in them we see that the Babylonians had developed a sophisticated system of divination based on the positions of the celestial objects. In looking at the night sky, they were seeking knowledge of the future. Astrology had been born.

The basic idea of astrology is that the state of the night sky has a two-fold influence over Earth. First, it influences natural events. Second, it shapes and influences our personalities. By carefully studying these associations, astrologers sought first to understand those links and then to use that knowledge to predict future events.

Although there continue to be many today who derive something from reading their horoscopes, academically astrology is a discredited subject. In our modern scientific world we have developed instruments to measure all of the forces that we observe in nature, yet nothing has been found that could possibly be carrying an astrological influence from the planets. From this modern perspective, it is therefore tempting to dismiss the development of astrology as an error made by lesser minds than ours. But we must again remember that the intellects of the Babylonian astrologers were every bit as sharp as ours; all they lacked was knowledge.

For example, we now know that the wind blows because there is a difference in air pressure between two geographic locations,[38] but the Babylonians saw it differently. Their world view was based on that of the Ancient Sumerians, who had explained the forces of nature as being the embodiment of gods, and therefore subject to divine will. It was only natural for the Babylonians to try to interpret the will of the gods by observing nature, as we might watch the behaviour of another person to determine their mood.

By the time the neo-Babylonian Empire had absorbed nearby Chaldea around 650 BCE, the astronomers had clearly assumed the role of diviners. Astronomy and astrology had become one discipline, and would largely stay that way in Europe until the seventeenth century. Astrology was systematically practised as a way to keep the heavens and Earth in balance. Indeed, so widespread was the practice that to be called a Chaldean was to be called an astrologer.

In Babylon, the first new moon after the spring equinox marked the beginning of the new year. This first month was called Nisan and its arrival sparked the beginning of the greatest festival in the Babylonian calendar. Known as the Akitu, it was a largely astrological event designed to mirror the supposed gathering of the gods in heaven, and involved ceremonies in which the wise men would determine what the coming year held in store.

The Akitu lasted for twelve days, beginning with the recitation of various prayers and the Babylonian creation myth, the *Enûma Eliš*. On the eighth day, the gods were thought to assemble, and in so doing they began to determine the character of the year ahead. On the eleventh day, the gods reached their decisions, which were displayed in the positions of the planets. This was the day that, on Earth, the astrologers would present their interpretations of the sky, relating the divine message to the King and his people. Other diviners would inspect the entrails of sacrificed animals and make their prognostications.

As with the rise of religion, it is clear that astrology was already well developed by the time that writing was invented. So, like time-keeping and religion, astrology's roots must date back into prehistory. Indeed, the belief that the night sky presented omens – secret knowledge that could only be interpreted by individuals schooled in special arts – is entirely in keeping as a grandiose development of the secret sky societies of the hunter-gatherers. By the time of the Babylonians, astrology is also indistinguishable from religion. We can think of it as a kind of 'theory of everything', a set of inter-related ideas that seek to explain the way heaven and Earth work together. It is only when religion becomes concerned with life after death, as in Egypt with the weighing of the heart, that the two ideas begin to diverge and compete, as we will soon see.

By 650 BCE, the Babylonians had compiled centuries of astrological predictions into a series of seventy stone tablets. Known as the *Enûma Anu Enlil* (In the Days of Anu and Enlil), they were excavated in the nineteenth century from the Library of Ashurbanipal, in Ancient Nineveh, which is located outside modern Baghdad. The library served King

Ashurbanipal and contained many thousands of clay tablets. Among numerous other things, these told the stories of the Epic of Gilgamesh and the *Enûma Eliš*. The *Enûma Anu Enlil* is a vast reference work. It preserves 6,500–7,000 omens, which are tied to a wide variety of celestial and meteorological events. The idea is simple: you observe something in the sky – maybe a thin crescent moon approaching the bright planet Jupiter – and then look this up in the tablets to see what it means. In short, the *Enûma Anu Enlil* is the first astrological encyclopedia.

The first twenty-two tablets are dedicated to observations of the Moon, which they associate with the goddess Sin. The next fourteen tablets record solar phenomena, and relate these to their sun-god Samas. The following thirteen tablets detail various meteorological observations. And the remaining tablets list various planetary alignments. Of these seventy tablets, number 63 is particularly well known among scholars because it very clearly records twenty-one years' worth of astronomical information regarding the planet Venus.

After the Sun and Moon, Venus is the brightest celestial object, but it is not a constant presence. It orbits the Sun, more closely than the Earth, so it hardly ever appears in a dark night sky. Instead, as it swings through its 225-day orbit, it drifts in and out of visibility, spending a lot of its time too close to the Sun to be seen with the naked eye. Because of the way the Earth's orbital motion combines with Venus's, the planet follows an eighteen-month cycle of visibility, known to astronomers as an apparition.

Venus first appears in the evening sky for a month or so. Then, as its orbit carries it between the Earth and the Sun, it disappears from visibility for a number of weeks before reappearing as an unmistakably bright 'star' in the morning sky. This is why the planet is sometimes referred to as the evening star and the morning star. It lingers for another six weeks or so, but as Venus begins to travel behind the Sun it disappears from view again, this time for a much longer period of time. It does not return to the evening sky until fourteen months later, completing one apparition and starting another.

Tablet 63 of the *Enûma Anu Enlil* records twenty-one years of this cycle. Although the tablets themselves were created around 650 BCE, attempts to date the astronomical information indicate that it could refer to apparitions of Venus that took place some time during 1700–1550 BCE. This too suggests a much longer astronomical tradition of creating a corpus of knowledge that was passed down through the centuries.

The significance of the *Enûma Anu Enlil* tablets is that in these carefully made astronomical observations we clearly see analytical, rational, even scientific minds at work, regardless of the erroneous attempt to link them with earthly events. Also, by this time in history, religion has transformed from the animism of the hunter-gatherers into a philosophy known as 'microcosm–macrocosm', or 'as above so below' – the idea that earthly events are mirrored in the cosmos. And since the religion of the time said that humans had been created to do the gods' work on Earth, our ancestors needed to interpret the natural messages and act on them. In this way, heaven and Earth could be kept in balance. If they ignored the heavenly messages, then chaos would return.

The practice of astrology was therefore thought to balance heaven and Earth and lead to civil order, political stability and economic prosperity. These days, little has changed except that civil order, political stability and economic prosperity are thought to be interdependent, and hence we have largely dropped the idea of worshipping gods and nature for their help in achieving this balance.

Once the Babylonians had absorbed the way celestial cycles correlated to large-scale natural events like seasons, they began extending their observations to smaller and smaller events. They painstakingly recorded their observations of the night sky and correlated them to political and social events, looking for patterns. This drove them to make sophisticated astronomical computations, and they became the first recorded people in history to use mathematics in their analysis of nature.

In addition to their recording and analysis of the visibility of Venus, the Babylonians calculated the way the day length varied during the seasons, and presented methods for calculating the time for the rising and setting of the Moon.[39] Clay tablets recovered from the houses of astrologers who lived in the first millennium BCE show that they also started to rely on their calculations of the planetary movements rather than actual observations.[40] From these they would predict where the planets would be in the sky and use that information to make their earthly forecasts. This reliance on mathematics shows how advanced and confident they were in their astronomy.

★ ★ ★

The planets hold a special position in astrology because they do not rise in synchrony with the seasons like the stars, nor do they follow the months of the Moon. Instead they keep their own individual timetables, from fleet-footed Mercury to the positively stately Saturn, and all rates in between. The Babylonians associated each planet with a god, and their configuration in the sky was thought to be the principal way those gods communicated with us.

There are five planets that can be seen with the naked eye: Mercury, Venus, Mars, Jupiter and Saturn. Apart from the Sun and the Moon (and the comets, which we will encounter in chapter 7), the planets are the only other celestial objects that change position against the fixed backdrop of stars.

Yet, as they chase each other across the sky, they never wander from the zodiacal constellations. We now understand that this is because the Solar System condensed out of a spinning disc of material, which fixed the planets, including Earth, into more or less the same orbital plane. To the curious minds of the past, however, the fact that the planets shared the Sun and Moon's path through the sky must have provided 'evidence' of their importance.

Highlighting the religious importance of these celestial trajectories, the Babylonians associated Mercury with Nabu, the god of literacy and wisdom; Venus with Ishtar, goddess of love, sex and political power; Mars with Nergal, god of war and pestilence; Jupiter with Marduk, the head god of the Babylonian pantheon; and Saturn with Ninurta, god of hunting and farming. In this way, Babylonian astrology began to associate the planets with different qualities and personalities.

The man responsible for entrenching astrology in the classical world was the second-century CE astronomer Claudius Ptolemy. He had already written his summary of astronomical knowledge in *Almagest*, and for his next work he wrote a defence and explanation of astrology known as the *Tetrabiblos*. It became a hugely influential piece of work, still referred to by astrological websites of the present day.

The book attempted to present a theory of everything based on Ptolemy's microcosm–macrocosm belief that 'most events of a general nature draw their causes from the enveloping heavens'.[41] Although he was not the instigator of all the ideas in the book, he synthesised the general trends of the time into a cohesive whole and made one important shift in emphasis. He championed the idea that astrology was a natural explanation for both the forces of nature and human personality rather than a supernatural signposting system for the gods, reinforcing it as separate from religion.

In the book, Ptolemy first draws the distinction between astronomy and astrology. The former, he writes, is designed to discover the celestial cycles, the risings and the settings of the various celestial objects and their movements. The latter then investigates the changes that these movements bring about on Earth. In other words, astrology is applied astronomy. He then sets about presenting his evidence for thinking this.

As everyone knows, the Sun has the greatest influence on the Earth's daily and seasonal cycles. It is responsible for day and night, and as the cave dwellers and megalith builders of prehistory discovered, its daily range of altitudes in the sky determines the seasons. Ptolemy pointed out the natural cycles that follow the Moon, such as tides and the behaviour of animals, and concluded that the Sun's dominant effects are modulated by the lunar and planetary positions and that these are transmitted to Earth by the weather patterns in our atmosphere. He goes on to state that, since it is clear that the seasonal weather patterns affect the quality and quantity of crops, plants and animals, it is entirely reasonable that they help determine human growth and personality too. By his theory, the 'temperament' of the very moment in which the individual is born is imprinted inside the person, and this temperament is determined by the position of the planets.

What is notable about Ptolemy's work is that he is proposing a strictly physical link between the planets and the air in Earth's atmosphere. He has taken the idea of godly intervention out of the equation. To do this, he drew on the work of the first Greek philosophers, known today as the pre-Socratic philosophers. They lived in the Greek coastal city of Miletus, in Anatolia (modern Turkey), and they embraced the ideas of microcosm–macrocosm. As a result, their ruminations explored the twin themes of nature (macrocosm) and human nature (microcosm), and began to separate the contemplation of nature from religion. They did this by looking for purely physical connections between things rather than attributing natural phenomena to the will of the gods.

Thales of Miletus is often cited as the father of Ancient Greek philosophy. He advanced the idea that both Earth and the cosmos were made of the same substance.[42] He suggested that the most fundamental form of matter was water and that everything else from rocks to air to fire was also water, just converted into different forms or phases. His contemporary, Anaximander, suggested that a separate, primal substance existed that embodied all four properties of ordinary matter: hot and cold, dry and moist. Ordinary matter then condensed out of the primal substance when these various contradictory properties were separated. A third member of the Miletus school, Anaximenes, returned to the idea of a known primary substance but chose this to be air, which could then condense into water and earth or thin into clouds and fire. By the time of Pythagoras in the sixth century BCE, the idea of the four classical elements – earth, water, air and fire – and their associated properties – hot, cold, dry, moist – was locked in. Each element was associated with two properties: earth is cold and dry, water is cold and moist, air is hot and moist, fire is hot and dry.

Pythagoras also wrote about the immortality of the human soul, or as he referred to it: the psyche. This had important consequences for astrology, as it proposed a link between the astronomical and the psychological realms; an idea that was subsequently taken up by Plato. Born in Athens in the late fifth century, Plato was a student of Socrates. What set the Socratic school of philosophers apart was that they brought a discussion of right and wrong human conduct into their contemplation of nature. Because of this, they are referred to as the moral philosophers.

Plato's theory of the psyche was that it originated in the heavens as part of a larger 'world soul'. Once united with a body, it provided the animating force to make it live. Thus, whereas animism postulated that everything had a soul, in Plato's view only living things possessed one. Since he believed that the human psyche was part of a larger 'world soul', which existed before the cosmos came into being, it meant that the whole universe was in fact a living thing. This allowed the moral philosophers to propose that the living universe would have a moral code, and this should be used to judge human behaviour. It also justified an astrological link between planets and personalities, because the psyche originated in the heavens.

Ptolemy used all of these ideas in his theory of astrology. He imagined that the Earth's atmosphere extended upwards past the planets, all the way to the stars. The Earth was the source of wetness in the cosmos, the Sun was the source of heat, and the stars were cold and barren. Each planet then derived its unique characteristic from its distance away from these sources. The considered order of the classical planets was the Earth, the Moon, Mercury, Venus, the Sun, Mars, Jupiter, Saturn, and the stars. Hence, Mars was said to be hot because it was next to the Sun, and quite dry because it was far from the Earth. This combination made it a choleric influence, so engendered leadership qualities and ambition. Jupiter, being slightly further away from the Sun and the Earth, was the ruler of sanguine individuals, who are outgoing and active. Saturn was the furthest planet from both the Sun and the Earth, thus it was deprived of heat and moisture and rendered melancholic, introverted and subject to anxieties. The final personality type in this scheme is the phlegmatic individual, ruled by the warm, moist planet of Venus, which sits comfortably between the Sun and the Earth. These individuals tend to be relaxed and easy-going.

So the theory arose that our personalities are set by the combination of these influences in their various strengths, as determined by the myriad configurations that the planets can assume. In presenting his ideas, Ptolemy admitted that astrology is often prone to error, but he put this down in some measure to inexperience on the part of the practitioners.

Ptolemy also explained how the planets could be used to diagnose and treat illness. Again, this idea probably has its roots in Babylonian astrology. In the late fourth century BCE, a Babylonian man named Iqīšâ lived in the city of Uruk. Excavations of his house have shown that he was a mašmaššu priest, an occupation that is often translated as exorcist but could be more sympathetically translated as physician. One of the principal jobs of a mašmaššu was to cure people of their maladies, which were often thought to be brought about by ghosts or invading spirits. The process of removing them was achieved by the patient performing rituals and consuming specially prepared food or drink, the forerunner of medicine.

Iqīšâ owned more than thirty scholarly tablets – the Babylonian equivalent of textbooks – some of which described how to prepare the curative foods needed in medicine. Notably their recipes depended upon the day of the year rather than the patient's symptoms. According to the tablets, each day of the year was associated with a position in one of the twelve signs of the zodiac. For example, day five of the fourth Babylonian month was associated with Taurus. To treat a patient on that day meant salving them with bull's blood, feeding them with bull fat, and fumigating their house with burning bull hair. If the treatment had occurred on the second day of the month, then the procedure would have been repeated except that, since the day was associated with Capricorn, the blood, fat and hair of a goat would have been used.

On the eighth day, however, which is associated with Leo, the system runs into problems. The trouble is that lions would have been extremely rare in Babylon. Archaeologist John Steele of Brown University, Rhode Island, suggests that either this prescription was a theoretically perfect system that could not be realised in practice, or the astrological associations were merely figurative and referred to common plants and herbs. If this is the case, then the astrological names were either used to keep this knowledge secret, or the Babylonians believed in correspondences between the various plants and the constellations.

At the same time in Greece, the physician Hippocrates was popularising a clearly related system of medicine based on the ideas of microcosm–macrocosm. He thought that since the year is a balance of the four seasons, and the world is made of four elements, so the body must be a balance of four substances. He referred to these as humours. It is possible that the idea of humours originated centuries earlier in Ancient Egypt, but it was Hippocrates who moulded it into a theory of medicine that we remember today. In his book *On the Nature of Man*, the humours are listed as blood, phlegm, yellow bile and black bile. The basic idea was that illness and disease occur due to deficiencies or excesses in these fluids. It therefore dispensed with the Babylonian idea of ghosts and spirits, but retained the notion that the imbalances could be traced to non-favourable alignments of the planets.

This was thought to occur because, just as the planets were associated with four personality types – melancholic, choleric, phlegmatic and sanguine – so too were the bodily fluids. Corresponding to no known bodily fluid today, black bile was supposed to issue from the spleen and drove the tendency to sadness and melancholy; it corresponded to the Earth element. Yellow bile was the equivalent of fire and came from the liver; it controlled irritability and choleric behaviour. Phlegm corresponded to water; it derived from the head and led to rational, phlegmatic behaviour. Finally, blood came from the heart and was the equivalent of air, engendering feelings of optimistic, sanguine behaviour. Balance between the humours could be restored in a number of ways. One was by eating the foods designed to balance them again.

Whereas the planets were thought to influence our temperament through the humours, the signs of the zodiac were thought to relate to different parts of the body. According to these ideas, the body is divided into twelve zones, each one linked to one of the zodiacal constellations. For example, Aries was the head and Pisces were the feet. Some difficult to interpret Babylonian tablets indicate that this idea could have taken form in Mesopotamia circa 500 BCE.[43]

Regardless of the completely apocryphal nature of these connections, Ptolemy's synthesis of astrological wisdom was so persuasive that our supposed connection to the planets was taken as read for the whole of the classical period. Belief then went into decline for more than four hundred years, beginning in the sixth century as the great Roman Empire crumbled and the bulk of Europe split into warring kingdoms. So began the

Middle Ages. The classical traditions of culture, art and science were largely trampled underfoot, lost and forgotten as mere survival became the order of the day. But astrology persisted nonetheless, as its ideas were kept alive in the Islamic world.

★　★　★

In the Middle East, Islam was born in the seventh century under the unifying influence of Mohammad, who is regarded by Muslims as the final prophet sent to Earth by God. A wave of expansion followed, which resulted in an Islamic empire that stretched from Spain and Portugal in the west, along the Mediterranean coast of North Africa, and through the Middle East to the borders of China and the Indian subcontinent. In the expansion, classic texts were translated into Arabic, incorporated into their academic tradition and expanded upon.

In the eighth century, under Caliph al-Mamum al Rashid, the world's first state-sponsored astronomical observatory was built in Baghdad. It began a trend that extended across the Muslim world. In the tenth century, Abd al-Rahman al-Sufi produced the *Book of Fixed Stars* (*kitabsuwar al-kawakib*), which took the star catalogue of Ptolemy's *Almagest* and described the constellations in Arabic terms. Many of the star names we use today derive from this very book. Examples include Algol (the demon – possibly named because it is the only star to visibly change its brightness every few days); Deneb, the tail (of Cygnus); and Rigel, the foot (of Orion). The names found their way into the western world in the latter centuries of the Middle Ages, from around the tenth century to the thirteenth.

The frontispiece from *Utriusque Coṣmi Historia* by Robert Fludd (1574–1637) depicts humankind as a microcosm within the macrocosm of the Universe. (AF Fotografie/Alamy)

One of the most important factors in this 'transmission' was the work of the Toledo School of Translators in the twelfth and thirteenth centuries. These were a group of scholars in the Spanish city of Toledo, which had been part of the Moorish empire but had passed back into Spanish hands in 1085.

Rather than despoil the libraries, the city's authorities protected the Muslim communities and cherished their books. The school translated many lost manuscripts from classical times, including Ptolemy's *Almagest*, reinvigorating medieval thinking and stimulating a huge interest in classical knowledge. This helped to ignite the Renaissance in Europe and led to a revival of interest in the ideas of microcosm–macrocosm and astrology.

Ptolemy's *Tetrabiblos* was translated into Latin in 1138, and had a profound effect not just on the revival of astrology but on medicine too, where the old astrological ideas were enthusiastically taken up. It became accepted that a correct medical diagnosis required analysis of the alignment of the planets and constellations in order to know which of the four humours were out of balance, and in which part of the body this imbalance was occurring. To locate the planets quickly, doctors took to carrying folding almanacs they could consult.

Although these almanacs must have been widespread, examples are now rare. Only thirty or so are known in the world. A particularly handsome one from the fifteenth century is owned by the Wellcome Library in London.[44] Written in Latin, it combines a calendar and astrological tables, all folded like a map. Once folded, the almanac is no larger than the palm of a hand, and fits into an ornate green and pink silk binding that then attaches to a person's belt. The centrepiece of this and other almanacs is the drawing of 'zodiac man': a representation of the human figure with the various signs of the zodiac superimposed.

The practice of bleeding a patient was designed to release the excessive humours that were causing the illness. To determine where to make the cut, a physician just had to consult the almanac. Once the unbalanced humour had been identified, the renaissance medicine man would look to see where the ruling planet was in the sky. Having determined in which sign of the zodiac it was located, they would then consult the zodiac man diagram to check which part of the body this corresponded to, and that was where the cut was made.

Outside medicine, astrologers were also becoming the therapists of their day, advising clients on the best way to make sense of their lives. This became known as electional astrology. Typically a person would go to an astrologer when they had an important decision to make. For example, should I ask the person of my dreams to marry me? Should I conceive a child? Should I attack my enemy? The astrologer would construct the client's horoscope at the moment of their birth, then look at the current position of the planets, and advise whether or not it was a propitious time for action.

The conviction that the stars could directly influence our lives was so strong at this time that celestial themes pervaded everyday discourse. Astrology reached its peak in England in the seventeenth century, when the recently invented printing press allowed cheap almanacs to flood the market place. These publications offered everything from sensible practical advice such as the phases of the Moon, which allowed people to plan their outdoor activities, to 'self-help' advice on how to live better lives in the coming year, based on the positions of the planets. In 1622, one almanac praised itself as containing:

Wit, learning, order, elegance of phrase,
Health, and the art to lengthen out our days,

Philosophy, physic, and poesie,
All this, and more, is in this book to see.

In the 1660s, a total of 400,000 copies of these almanacs were in circulation in England, which translates into about one in three families buying one.[45] This placed the discipline on a collision course with the Church of England. There had always been an uneasy relationship between Christianity and astrology, not least due to the difficulty of reconciling a belief in free will with the notion that human actions are determined by the position of the planets. Indeed, in the Book of Leviticus – one of the oldest parts of the Bible – God's followers are expressly banned from practising astrology or any other form of divination.

But the Bible is inconsistent on this matter. In the New Testament, the Gospel of Matthew describes the birth of Jesus, stating that three 'wise men' or kings were alerted to his arrival and guided to the crib by the star of Bethlehem that God had placed in the heavens. Clearly, this was an astrological event, and the original word used for the wise men was magi, the plural of magus. This is a provocative term. In its original Greek form, it referred to the followers of Zoroaster, the originator of one of the first monotheistic religions, which established itself in what is now Iran. Indeed, the biblical magi are said to come from the east. In later centuries, however, the term came to mean astrologer or a practitioner of the occult arts. The word magic is derived from this root.

The reframing of the magi as kings started in the third century, when theologians pointed out that if they were royal then their appearance at Jesus's crib could fulfil a prophesy made in the 72nd Psalm that all kings would worship him. But it wasn't until the King James Bible of 1611 that they were enshrined as 'wise men' rather than astrologers. By then, astrology had gripped the general population in a way that threatened the authority of the Church in Britain, and something had to be done.

King James of England ordered the Bible to be translated into English from Latin so that the common folk could finally read the holy book rather than rely solely on the word of clerics. The translators changed every single occurrence of 'magus' to 'magician' or 'sorcerer' to help demonise their practice. There was just one exception: the three magi who visit Jesus to worship him are transformed into 'wise men'.

Christianity wasn't the only religion to have problems with astrology. Ibn Qayyim al-Jawziyyah, a fourteenth-century Islamic theologian, took the subject to task along with other occult practices in a 200-page excoriation called *Miftah Dār al-Saʿādah*. He thought the universe around us was God's gift and a sign of His cosmic perfection in bringing order out of chaos. It was folly for a human to think that they could understand even the smallest scrap of it.[46]

Those who believed that human personalities came from the stars were, according to al-Jawziyyah, 'the most ignorant of people, the most in error and the furthest from humanity'. Indeed, he argues that astrologers are worse people than infidels, by which he means Christians. One of his quirkiest arguments is to suggest that if stars and planets had some form of intelligence or wit they would surely exercise it to leave their prescribed orbits. Since they do not, he says, they must surely be bound by the Almighty's will. A more straightforward argument comes a little later when he asks why twins born mere moments apart can have different personalities.

After centuries of dominance, astrology collapsed as an academic discipline in the latter seventeenth century. In England, at least, this was partly because of astrology's role in the Civil War. Waged between 1642 and 1651, it saw the execution of King Charles I and the establishment of Oliver Cromwell as Lord Protector of England. This flirtation with being a republic came to an end in 1660 when Charles I's son was invited back from exile and crowned Charles II. During the war, astrologers on both sides were busy making predictions of victory that were apparently written in the stars, but those on the side of the republicans were by far the most vociferous and successful – after all, their side won. Mirroring this propaganda war, the poet John Milton described the war in celestial terms:

Among the constellations war were sprung,
Two planets rushing from aspect malign
Of fiercest opposition in mid sky,
Should combat, and their jarring spheres confound.

Upon the restoration of the monarchy, astrology fell into disrepute in England because of its association with the predictions of defeat for the Royalists. But times were changing across Europe too. The philosophy of microcosm–macrocosm was being replaced by empiricism, which championed the careful observation and measurement of nature.

Without its underlying philosophy to support it, astrology was doomed. Yet even today the English language is shot through with references to the night sky that come from astrology. Once, the word 'luminary' was used by the astrologers to mean either the Sun or the Moon, the principal astrological influences. Now it means a person who can inspire or influence others because of their great knowledge or standing. We continue to refer to good luck as happening when 'the stars are aligned', and we continue to describe people in astrological terms. They can be saturnine, meaning slow and gloomy; mercurial, meaning unpredictable; jovial, meaning cheerful and friendly; or martial, meaning warlike.

When astrology collapsed in the seventeenth century, weakened by the attacks from the Church, it brought us to a watershed moment in our relationship with the night sky – one from which some philosophers argue we are still reeling. But to see the inevitability of this sea change, we have to step back in time once more – to the classical philosophers of the sixth century BCE – and follow a parallel train of thought in the attempt to extract meaning from the night sky. And although it, too, has been superseded in modern times, it gave rise to one of the most enduring images of a Christian heaven: the cherub playing a harp.

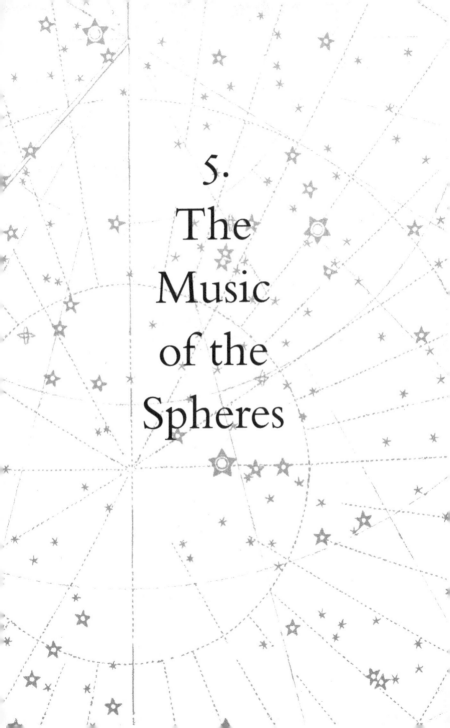

5.
The
Music
of the
Spheres

In the sixth century BCE, the great philosopher Pythagoras established a secret school dedicated to intellectualism. The Pythagorean brotherhood believed that one of life's goals was to purify body and soul so that our eternal spirit could return to the heavens when our time on Earth was done. In that quest, the more one knew about the cosmic perfection of the night sky, the more ordered one's life and thinking became, and the better prepared one's spirit was to return to the stars after death. Taking this to its conclusion, Pythagoras thought that our study of the perfect cosmos would allow us to become perfect ourselves, and ultimately indistinguishable from gods. But how can we truly understand the heavens when all we can do is look in awe and wonder at their endless circulations?

According to legend, Pythagoras had his epiphany while passing a blacksmith's shop, where he was stopped in his tracks by the sound of hammers against anvils. Each of the smiths' hammers was producing a different musical note, and as they fell some pairs combined into pleasant consonance, while others were unpleasantly dissonant. Seized by curiosity, Pythagoras rushed into the shop to investigate.

Once inside, he discovered that the different sounds were produced by hammers of different weight. He made the smiths strike the hammers in their various combinations to see which pairs produced consonance and which dissonance.

In this way, Pythagoras identified the most consonant musical intervals: the octave, the perfect fifth, and the perfect fourth. Upon examining the hammers, he was staggered to discover that strikingly simple mathematical relationships existed between the hammers that produced these intervals. For the octave, one hammer was half the weight of the other; for the perfect fifth, one hammer was two-thirds the weight of the other; and the perfect fourth was produced when one hammer was three-quarters the weight of the other.

The trouble with the story is that it is untrue.

Many ideas are attributed to Pythagoras. As none of his works survive today in written form, however, we can only learn of his thinking, his methods and his conclusions through the works of others. The earliest known version of the hammers story is found in the second-century writings of Nicomachus, a mathematician from Roman Syria (modern Jordan). The problem with it is that no matter how heavy the hammers are, it doesn't affect the pitch. Change the hammers and anvils to plucked strings of differing lengths, however, and these mathematical relationships can be demonstrated. Indeed, this is exactly what happens in the next, more believable part of the story, where Pythagoras is said to have rushed home to begin his own experiments.

He constructed an instrument known as a monochord, a single string suspended above a soundbox. The string is supported at one end by a fixed bridge and at the other by a tuning peg that allows the string to be tightened. In between is a movable bridge that allows the string length to be altered, and hence the pitch of the note it produces when plucked.

Compared to the pitch of the open string, placing the movable bridge halfway along gives the octave; placing it two-thirds of the way along the string gives the perfect fifth, and three-quarters of the way along gives the perfect fourth. That such simple ratios as these – 1:2, 2:3 and 3:4 – gave the most consonant musical intervals was a revelation to Pythagoras. Indeed, the discovery seemed to prove a principle that he and his followers had rested their philosophy upon: that nature was the embodiment of numerical relationships.

At school we're taught about Pythagoras's theorem concerning right-angled triangles. It says that if you square the lengths of each of the two smaller sides, their combined area is equal to the squared length of the longest side, the hypotenuse. As well as being a neat trick with triangles, it implies something extraordinary about nature: that shapes and structures are the physical reality of otherwise hidden numerical relationships. We can investigate these numerical relationships if we translate nature into numbers. This is what Pythagoras did with the monochord when he measured the position of the bridge along the string.

The Pythagorean belief in the primary importance of numbers led to the phrase that now sums up his philosophy: 'all is number'. This is the idea that without numerical relationships to define its shape and size, an object simply cannot exist. In other words, the abstract realm of numbers is more fundamental than the physical. This belief now underpins all modern science: that nature is understandable by converting it to numbers through measurement, and then analysing those numbers to find the mathematical relationships, which we now call the laws of physics.

Pythagoras's discoveries with the monochord led him to believe that music was a unifying system that could be applied across nature. He and his followers believed that odd and even numbers were fundamentally different things. Odd numbers represented limitations and moderation and were therefore seen as good, while even numbers represented unlimited things and were seen as bad because having no limits on one's behaviour can lead to ruin. The Pythagorean school postulated that only music had the power to bring the two together and mould them into something beautiful, something harmonious. Their reasoning was that each of the ratios that described the consonant intervals contained an odd and an even number: 3:4 for the perfect fourth, 2:3 for the perfect fifth, 1:2 for the octave. In other words, music brought order and balance to what the Pythagoreans saw as diametrically opposed quantities.

The concept of balancing opposites abounds throughout Pythagorean philosophy. In the work of one of his followers, Philolaus, the cosmos and everything in it can be divided into limiters and the unlimited. The celestial objects with their definite shape and structure are limiters, while the vast unbounded realms of space and time are the unlimited. And to bring them together, Pythagoras looked once more to music.

He noted that a sound becomes musical only when it is preceded or followed by another, or produced simultaneously. Our ear judges the difference between the two notes – the interval – and decides whether it finds it consonant and pleasant or dissonant and unpleasant. The interval can then be written mathematically. After pondering how this might be applied to the heavens, he proposed the *musica universalis*; the music of the spheres. Just as the various notes in the scale are achieved by placing the monochord's movable bridge at different distances along the string, so the various celestial objects are all positioned at different distances from Earth. Thus Pythagoras imagined each planet emitted a note tuned to its distance from Earth, and together those rang out in a grand universal harmony.

Pythagoras, like many others of his time, imagined that Earth was the fixed centre of the cosmos and that everything revolved around our world. As the stars are the furthest things from Earth, so Pythagoras equated their distance to an octave. He then represented each of the other celestial objects as notes on the scale, separated by a musical interval. In western music today, we tend to use seven-note scales, where the notes are separated by semitones or tones. A tone is composed of two semitones, and the pattern of semitones and tones determines whether the scale is major or minor.

Pythagoras's scale had a note to represent each of the celestial objects: the Earth, the Moon, Mercury, Venus, the Sun, Mars, Jupiter, Saturn and the stars. This made it an eight-note scale with the pattern of intervals being tone, semitone, semitone, three semitones, tone, semitone, semitone, semitone. Together these intervals add up to the six tones (or twelve semitones) of an octave. They give a restless minor-key feel to the scale, and if all those notes were to ring together, the result would be anything but harmonious.

And so began an odyssey that spanned around two thousand years. The concept of the music of the spheres and the search for universal harmony was explored and extended, honed and decorated, before finally being discarded as science proper grew to prominence. To modern eyes, we see that the error lay in believing that the mathematical equivalent of musical intervals was the only numerical relationship that differently spaced objects can have in nature. But to Pythagoras, music was the first natural thing to be captured in the form of numbers, and he assumed the universe would have been made to mirror such beautiful natural harmony.

This idea was enshrined two centuries later in Plato's *Republic*. Plato embraced many of Pythagoras's teachings, and in his masterwork he presented the music of the spheres as the grand organising principle behind the movements of the night sky. Just in case that sounded a bit dry, he also merged it with his theory of the soul and the idea of a judgemental God deciding our fates, and then disguised the whole thing as a myth in which a tremendous battle is raging.

The warrior Er is fighting for his life against impossible odds. When the inevitable happens, his body falls to the ground; yet in the days that follow it shows no sign of decomposing. On the twelfth day, having been moved to the funeral pyre, Er opens his eyes. Magically his soul has been returned to his body, his life has resumed and he has a story to tell.

He's been to the afterlife, and has seen the links between heaven and Earth. He's witnessed how souls are judged and either rewarded or punished for their earthly behaviour. Those souls who have committed petty offences are taken to the underworld and made to atone tenfold for their crimes. The good ascend and are rewarded. Then both groups are brought back together and each soul is asked to choose a new life so that they may be reincarnated and returned to Earth.

Prior to their return, they are shown the true arrangement of the cosmos. At its centre lies a spindle, held by Ananke, the Greek deity of inevitability, compulsion and necessity. Around the spindle are the orbits of the planets, all kept in motion by the three Fates, who sit in attendance beside Ananke. On each orbit sits a Siren, singing a single pure note depending on how fast the Fates are turning the orbit. Together those notes blend to form the serene harmony of the music of the spheres.

The reincarnated souls are then required to drink from the river of forgetfulness. As they fall asleep, all memory of their judgement is lost, and instead of waking in the heavens, they reincarnate back on Earth. But this is not Er's fate; instead of being judged, he is told to observe and report his new knowledge back to humankind.

Written in this way, it is easy to forget that the idea of the *musica universalis* was a well-motivated, serious attempt at investigating nature, and although it was developed long before the recognition of the scientific method, it bears most of its hallmarks.

What we now call science was codified during the sixteenth and seventeenth centuries. It is based on the development of clear mathematical relationships between a system of natural objects or phenomena. These relationships are used to build a hypothesis that seeks to describe the system as a whole. The hypothesis is then used to make predictions about how that system will behave in an as-yet-unseen situation, and these predictions are tested by observation or measurement. If the predictions are shown to be true, the hypothesis is renamed a theory. If the predictions fail, the hypothesis is adjusted and the sequence starts again. This is the scientific method in a nutshell.

In the case of the music of the spheres, the mathematical relationships were the ones that Pythagoras had shown relate to the consonant musical intervals. The hypothesis was that the celestial objects are located at distances corresponding to notes on a musical scale. And the testable predictions can be thought of as the calculation of which note each planet emits.

In the earliest writings on the subject, Pythagoras and his followers concerned themselves only with intervals, the relationship or distance between the notes that represent the celestial objects. For example, Pythagoras is said to have calculated that the Earth and the Moon are 126,000 stadia apart, where a stadium is 625 paces. He made this distance to be the celestial equivalent of a tone. While this is certainly a good start, the question that really needs answering is: what pitch does the Moon emit? The first recorded attempt to assign actual note values to the celestial objects comes from Nicomachus in his book, *Manual of Harmonics*.

His planetary scale is based on a seven-note sequence that starts on D and moves downwards through the natural notes, except for B, which is flattened. Knowing that the faster a string vibrates, the higher the pitch it emits, Nicomachus assigned the fastest moving celestial objects to the highest notes. So in this case, the Moon was D.

But this hypothesis fails the test because if they were to be played together, the resulting chord would sound highly dissonant.[47] Perhaps this explains why others turned to one of the fundamental building blocks of Greek music theory: the Greater Perfect System.

This was a sequence of notes that covered two octaves. Within the sequence, the pitch of certain notes was fixed, while others could be tuned by ear according to the musician's personal taste. The celestial objects were assigned to the fixed notes, which were either a perfect fourth or a tone apart, and this put greater harmonic distance between them, allowing the resultant chord to ring somewhat more consonantly.

In a different system again, Ptolemy, the author of *Almagest*, not only included the celestial objects but split the Earth into its four classical elements – earth, water, air and fire, and suggested these too had musical significance. In *Timaeus*, Plato wrote some convoluted sentences that academics have interpreted as an interval sequence for the planets. These are based on the perfect Pythagorean intervals, so that the Moon to the Sun is an octave; the Sun to Venus is a fifth; Venus to Mercury is a fourth; Mercury to Mars is an octave; Mars to Jupiter is a tone (i.e. the distance between a fourth and a fifth); and Jupiter to Saturn is an octave and a fifth. Extending over more than five octaves, it does have a mostly consonant aspect to it.

On first encountering these various ideas, it's tempting to dismiss them all as guesswork, but really they are just competing hypotheses that must be tested. Even today, modern astronomers steeped in the scientific method develop many competing ideas to explain puzzling observations and then make new observations to decide which is right.[48]

There was something that struck the classical philosophers as beautiful and therefore right about the music of the spheres, and this entrenched their belief in it – even though it was becoming clear that the idea was too simplistic. For example, astronomers were clearly seeing that the movement of the planets was more complicated than first thought. The planets do not move at constant speeds night after night. Instead they speed up and slow down, sometimes decelerating so much that they double back on themselves before resuming their forward motion. None of this behaviour was captured in the various scales and schemes of celestial harmony. By today's scientific standards it should have been the death knell for the idea, but then science was in its infancy, and the importance of proof had yet to be fully grasped.

Plato even cautioned against trying to test the music of the spheres, saying astronomers and musicians were taking their subjects too literally. He thought that true knowledge existed only as abstract, mathematical relationships that could never be checked by experimentation or observation. In the *Republic*, he wrote, 'All the magnificence of the heavens is but the embroidery of a copy which falls far short of the divine Original, and teaches nothing about the absolute harmonies or motions of things.'

In other words, no matter how closely we look at the universe, or how deeply we listen to music, our perfect mathematical theories will never translate into physical reality because matter somehow corrupts them. This became a cornerstone of classical thinking. The works of fourth-century BCE philosopher Aristotle taught that the heavens were perfect and the Earth was corruptible. This led to the idea that the celestial objects were made of a perfect substance: a fifth element called quintessence or aether. It made perfectly spherical celestial objects, which travelled in perfectly circular trajectories, and sang perfectly harmonious music.

The idea of the music of the spheres endured throughout the classical period but languished without further development until the sixth century, and the brink of the Middle Ages. In this time of transition, only a handful of individuals clung to the old ways of knowledge. One in particular was the philosopher Anicius Boethius. Born into the ruling classes in Rome in 477, he became one of the most influential medieval philosophers. He detailed the previous thousand years of musical discussion in his book, *Formation of Music*, which became the standard musical reference for the next millennium. It transferred the idea of the music of the spheres from the philosophers to the music theorists, and set the stage for another attempt to provide a plausible explanation for astrology.

In his book, Boethius grouped music into three categories. Of these, only one is now recognisable as music to us: *musica instrumentalis*. This is the music produced by humans, either with instruments or voices, and according to Boethius is the lowest form of music because it is an inherently imperfect attempt to capture the purity of music that can be naturally found in the heavens. Not only that, but the person who plays the instrument is the lowest form of musician because they don't necessarily understand what they are doing – they are little better than a device reading a manuscript and mechanically producing a sound. The composer is on a slightly higher intellectual level because they imagined the music in the first place, but even this is no guarantee that they understood what they were doing.

As extraordinary as the concept might seem today, Boethius defined the true musician as the person who sits in an audience, listens to the music and understands it. This elevation of music to an intellectual discipline leads to his other two categories, where music is used entirely intellectually rather than for the pleasure a good tune can give us. The pinnacle is *musica mundana*, the music of the spheres, and just below that is *musica humana*, the inner music of the human body.

In discussing *musica mundana*, Boethius gives us the resolution of a debate that had dogged the subject since its inception: whether the universal harmony was audible or not.

In originating the idea, Pythagoras is said to have claimed to be able to hear it. Yet clearly none of the rest of us can. The supporters of Pythagoras suggested that this was because we have become so accustomed to the music that we are no longer continuously aware of it. Others thought the music was entirely theoretical. Even if the mathematical ratios that produced consonant musical intervals were applicable to the distances of the planets, that did not mean that actual notes were produced.

Aristotle reasoned that the size and speed of the celestial objects should produce mighty sounds, which would carry immense force. So, noting that excessive noise could shatter solid objects, he reasoned that the continued existence of the Earth argued against actual music coming from the universe. In turn, Cicero felt that human ears were simply not equipped to hear it – in the same way that our eyes are not equipped to look directly at the blinding light from the Sun.

In *Formation of Music*, Boethius reasoned that the sounds were real yet inaudible. Although we cannot hear the music of the spheres, he proposed that nature resonates to the song. According to Boethius, it is the very reason for the changing of the seasons. This marks a significant change in the role of the music of the spheres. Originally it just explained the distances of the planets. Now, in effect, Boethius proposed that the music of the spheres was the astrological means through which the planets influence the Earth.

The idea that the seasons were related to music goes back to the Ancient Orient. The Chinese identified a sequence of four notes – F, G, C and D – said to correspond to autumn, winter, spring and summer. According to one legend, the music master Wen of Cheng could change the seasons according to which two strings he plucked on his zither.

Boethius identified the music as coming from the heavens rather than a person, but the idea is essentially the same: music has the power to transform nature. And in Boethius's third type of music, the *musica humana*, the inner music of the human body, he discussed music's ability to transform us. 'Music,' he wrote, 'is so naturally united with us that we cannot be free from it even if we so desired.'

The idea dates back at least as far as the fifth century BCE, when Plato popularised the notion that music formed an integral part not just of the cosmos, but also of our psyche. To Plato music was essential preparation for the human soul. Through its consonance and dissonance, it can give us our first education in pleasure and pain, and how to respond correctly to these stimuli. He points out that in trying to get a restless baby to sleep, a parent will gently rock it and sing a lullaby, thus using rhythm and pitch as a means to stillness and silence. 'More than anything else,' he writes, 'rhythm and harmony find their way into the inmost soul and take strongest hold upon it.'

Rather like the justification of astrology that there are certain phenomena that are clearly relatable to astronomical configurations, such as high tides and the seasons, so there is justification for thinking about music as a primal force. We all know that music has the power to move us. Songs in major keys are often described as happy, while those in minor keys tend towards sadder emotions. The different types of keys are produced by simply choosing different sequences of notes. One pattern gives the major key, another gives the natural minor. Within the minor keys, there are two slightly different sequences that give us the harmonic minor and the melodic minor. There are other sequences we can choose that give us the so-called modes. And all of these have subtly different musical characteristics.

To the classical philosophers these subtleties determined their emotional impact. Plato even prescribed which scales should be listened to in order to facilitate a person's entry into their chosen profession. For example, he suggested soldiers should listen to Greek Dorian or Phrygian modes to make themselves stronger.

By applying the music of the spheres to such a wide variety of phenomena, Boethius opened the door to its popular acceptance. Throughout the Middle Ages and into the Renaissance, the idea of music as the link between the night sky and the human soul was widespread. In *The Merchant of Venice*, Shakespeare wrote:

Sit, Jessica. Look how the floor of heaven
Is thick inlaid with patens of bright gold.
There's not the smallest orb which thou behold'st
But in his motion like an angel sings,
Still quiring to the young-eyed cherubins;
Such harmony is in immortal souls;
But whilst this muddy vesture of decay
Doth grossly close it in, we cannot hear it.

The idea was developed to its fullest in the work of Franchino Gaffurio, an Italian music theorist and composer of the late fifteenth century. For the frontispiece of his 1496 book *Practica musicae*, he commissioned an extraordinary woodcut.

Down the centre of the page wriggles a three-headed serpent. At its tail (at the top of the page) sits Apollo on his throne in heaven. He is holding a lute and is flanked by cherubs with musical instruments of their own. At the serpent's heads is Earth, which is subdivided into the classical elements, earth, water, air and fire. Along its snaking body are the planets on the right-hand side, and the Greek muses on the left. Between the planets, the music of the spheres is notated as intervals in a sequence that builds into the natural minor scale, which consists of only natural notes starting on A. These notes are identified on the left-hand side by their names from the Ancient Greek Greater Perfect System – but Gaffurio doesn't stop there.

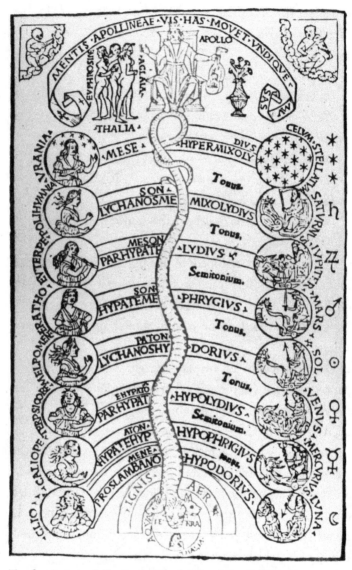

The frontispiece of *Practica musicae* by Franchino Gaffurio (1451–1522) depicts the concept of the music of the spheres by associating the planets with notes on a natural minor scale. (Charles Walker Collection/Alamy)

Instead of each planet being associated with a single note, each is given a unique scale. These scales are what we know today as the Medieval Church modes. Each one has a different musical characteristic, and their inclusion could well be an attempt to relate astrological properties to the planets through the emotional effect the various musical modes have on people.

It also reflects a change in the style of music that people were listening to in the Middle Ages. Music had transformed from single melody lines into something that embraced numerous melodies sounding together to give rich harmonies. Termed polyphony, it was seen as the realm of the professional composer, whereas the simpler monophonic songs, however virtuosic, were increasingly labelled as amateur folk music.

In polyphony, people were discovering a wealth of new moods and feelings that music could express, and in the work of Gaffurio this richness was absorbed into the music of the spheres. According to his diagram, instead of each planet producing a single note to give a single celestial chord, the planets sang according to their mode and these varying notes mingled in an extraordinary, ever changing polyphony that influenced the myriad events on Earth. In such vaulting ambition we see the fullest expression of the music of the spheres that history has to offer.

The rich complexity of such a system meant that rather than try to deduce the grand celestial polyphony note-for-note, most were content to think about the music of the spheres in a purely metaphorical manner. There was one man who was undaunted by the challenge. He was a sixteenth-century German mathematician who wanted to return to the strictly mathematical Pythagorean roots of the music of the spheres. Johannes Kepler believed that mathematics was a language that allowed complete precision – unlike words, which are always open to interpretation. As a devout Lutheran Protestant, Kepler thought of mathematics as the divine language of God. This meant the movements of the planets must at heart be mathematical, and therefore musical too because, according to Pythagoras, music was synonymous with numerical relationships.

In his search to transform the movement of the planets into music, Kepler began an intellectual endeavour that changed our relationship with the night sky for ever.

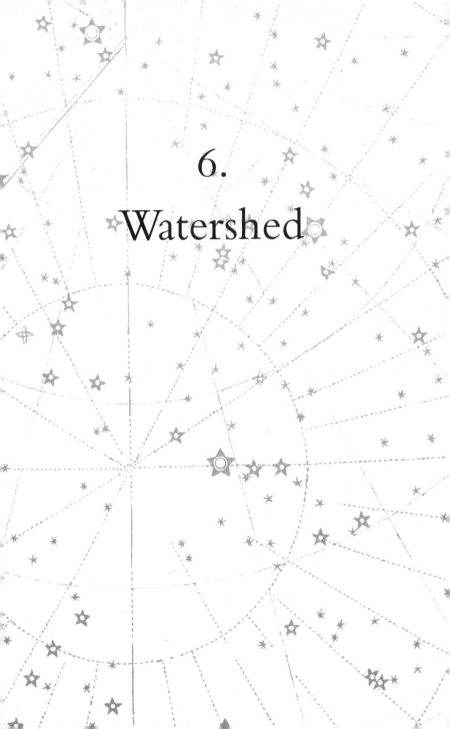

6.
Watershed

In the early sixteenth century, the motion of the planets was still explained in a way that dated back to the classical philosophers of the sixth century BCE. It relied on the idea of celestial spheres or orbs. These were a sequence of transparent, nested spheres made of the aether (the fifth element) that rotated inside one another, carrying the planets along with them.

In the earliest of these schemes, the fastest orb was the most distant: the sphere of the fixed stars. It travelled essentially a full circle once a day, moving from east to west. The classical planets, which included the Sun and the Moon, moved more slowly on their own individual orbs at their own individual paces, which explained why they moved westwards night after night against the backdrop of stars. But, as with the celestial harmony of the previous chapter, this simple scheme failed to explain the various accelerations, decelerations and outright backtracking of the planets. To address these deficiencies, Ptolemy had proposed various solutions in his book, *Almagest*.

For the backtracking of planets, he suggested that the shell of each celestial orb was thick enough to contain a smaller orb, called an epicycle, which rotated independently as the larger orb turned. The two motions together carried the planet on a spiralling path that would explain this occasional phenomenon.

For the speeding up and slowing down, Ptolemy proposed moving the Earth slightly away from the centre of the celestial spheres. This meant that, from our point of view, the planets would appear to speed up and slow down. But neither of Ptolemy's ideas accurately reproduced the observed planetary motion, and many scholars were also uncomfortable with the idea of an offset Earth because it contradicted Aristotle's theory that earth was the densest element and so our planet would naturally sit at the very centre of the cosmos. The theologians were unconvinced, too, as it made little sense that God would arrange the heavens around an empty point in space. Yet despite these unresolved issues, both Ptolemaic and Aristotelean ideas penetrated medieval and renaissance academic thinking.

The idea of celestial spheres found its way into the wider public consciousness because both Christian and Islamic theologians incorporated Ptolemy's astronomy into their view of the cosmos. They simply added God as the 'prime mover', whose omnipotent power turned the orbs in the first place. It was not a new idea that some form of intelligence was behind the movement of the celestial objects, but these new theologians included an unmoving sphere beyond the stars that served as God's dwelling place; hence the term 'seventh heaven', which means the actual divine realm rather than any one of the classical planet's spheres.

It was at the beginning of this debate, around CE 1000, that the association of Christian heaven with the night sky became a central pillar of the religion. Before that time the old English word 'Hefon' could simply mean the night sky. The word firmament was also adopted by Christian academics in the Middle Ages to describe the sphere of the fixed stars, which they saw as a solid dome and the boundary between the visible realms of the cosmos and the seventh heaven. In some versions of this idea, the stars themselves are holes in the firmament through which the pure white light of heaven shines.[49]

As popular as these ideas became, there were still those troubled by the shortcomings and incompatibilities of Ptolemy's model. One of those most troubled was the Catholic canon and astronomer Nicolaus Copernicus, who lived and worked in Frombork, Poland.

During the 1510s, Copernicus gathered tables of existing astronomical observations and then made observations of his own in order to draw up a new idea for how the heavens might be arranged. These ideas were presented in a document entitled *Commentariolus*, which he was at first careful to show only to a few friends, as his theory contradicted the teachings of both Ptolemy and Aristotle. Copernicus had drawn the conclusion that planetary movements could be better explained if the Sun was the centre of the cosmos, with the Earth just one of the other planets following an orbit of its own around the Sun. This seemed to provide a natural explanation for why the planets occasionally appear to backtrack in the sky.

Imagine all the planets on a circular racetrack centred on the Sun. Earth is in lane three. Inside it are Venus and Mercury; outside it are Mars, Jupiter and Saturn. Being on the inside tracks, Mercury and Venus are travelling faster than Earth. As they lap us, they are moving in the same direction, but by the time they reach the far side of the track, from our perspective they now appear to be travelling in the opposite direction, hence they appear to move away from the Sun and then change direction to head back towards it.

Mars, Jupiter and Saturn are on the outside tracks. They are moving more slowly and we are the ones doing the over-taking. As we slide past on the inside track, our line of sight shifts and the outer planets appear to move backwards across the sky from our perspective. Think of it like being on one train overtaking another. As we pass by, the slower train looks as if it is moving backwards when in reality we are both travelling in the same direction, just at different speeds.

The idea of Earth orbiting the Sun was daring but not new. Doubts about Earth's immobility had surfaced repeat-edly since the days of Ancient Greece. In the fifth century BCE, Philolaus proposed that day and night were produced by Earth orbiting an invisible, central point called the 'central fire' or 'Zeus's watchtower' once a day. It was this that made the more distant Sun, stars and the other celestial objects appear to cross the sky. Although it is not a heliocentric cosmos (with the Sun in the centre), it does correctly ascribe most of the movement of the night sky as being down to Earth's motion (even if it doesn't get that quite right either).

A fully heliocentric view was proposed by Aristarchus of Samos in the third century BCE. He identified the central fire as the Sun, and so moved it to the fixed centre of the cosmos, proposing that Earth spun on its axis to produce day and night. In the Islamic world during the Middle Ages, several astronomers had also favoured the idea that day and night were created by Earth spinning on its axis. Surviving correspondence from Iranian astronomer Abu Sa'id al-Sijzi in the eleventh century makes it clear that he worked from this assumption.[50]

For reasons lost to history, however, Ptolemy favoured the geocentric model. Perhaps it was because so much additional meaning had been hung on Earth's central position, such as Aristotle's theory of elemental density. Whatever the reason, it was the *Almagest* that became the standard work and hence its geocentric system that was taken up by the religions and cultures of the time – and this made Copernicus cautious about publishing his ideas.

Following his initial pamphlet, the Polish astronomer worked for decades to hone a complete version of his model that could reproduce the movement of the planets across the night sky in all their subtleties. Although he did not manage to do that, he was satisfied enough with his results to have them published just before he died in 1543 in a book called *De revolutionibus orbium coelestium* (On the Revolutions of the Heavenly Spheres). In his introduction, Copernicus explained that his motivation was to provide a more accurate theory of planetary motion to increase the precision of astronomical predictions and so allow a better calendar to be produced.

This was a major concern at the time. Catholic countries had been using the Julian calendar introduced by Julius Caesar in 46 BCE. This was a solar calendar that rested on the calculation that a year lasted 365.25 days. Most years of this calendar lasted for 365 days, but every fourth year was named 'leap' and included an extra day. It was by far the most accurate calendar of its time, but it contained a small error. The average length of a year is not 365.25 days but

365.24217. This meant the Julian calendar overestimated the length of a year by eleven minutes. In the decades following its introduction this error was hardly noticeable, but by the sixteenth century, the discrepancy had built up so much that the calendar and the seasons were around ten days apart. The spring equinox was taking place on 10 or 11 March instead of 21 March. This was important to the Church because it was interfering with setting the date of certain religious festivals. Something had to be done.

Roman Catholic astronomers across Europe began looking to the sky again, making new measurements of key moments such as the time of equinoxes and solstices, so that the length of the year could be more accurately calculated and a better system of leap years devised. Copernicus was asked to contribute to the effort, and the heliocentric system he proposed was subsequently used in 1551 by the German astronomer Erasmus Reinhold to compute the Prutenic Tables. This gave updated positions of the stars and planets at various times, and was one of the principal sources used in the construction of the new calendar.

Eventually it was agreed that every fourth year would be a leap year except in instances when it fell on a century year that was not divisible by four hundred. This meant that the years 1600 and 2000 were leap years, but that 1700, 1800 and 1900 were not. This small change meant that instead of there being a hundred leap years in every four hundred years, now there were just ninety-seven. On 24 February 1582, Pope Gregory announced the new calendar.

To pull society back in line with the movements of the night sky, ten days were to be added overnight to the old Julian calendar, which would then be replaced by the new Gregorian calendar.[51] This was set to happen at midnight on 4 October 1582, with the next morning officially becoming 15 October. But despite the practical advantages, religious rivalries meant that not everyone was keen to do this. Protestant Christian countries in particular were unwilling to follow a Roman Catholic lead.

It wasn't until almost two centuries later, with the Calendar (New Style) Act of 1750, that Great Britain and its colonies adjusted. In Sweden, they planned to make the reform gradually by excluding all leap years between 1700 and 1740. However, they somehow forgot to eliminate the leap years in 1704 and 1708 and this led to the abandonment of the plan entirely, so they had to add a day into 1712, making a double leap year, to get them back in step with the old style calendar. The confusion finally came to an end on 17 February 1753 when at midnight, the rest of the month was dropped and Sweden woke up to 1 March and the Gregorian calendar.

Although Copernicus's heliocentric system had been useful, it was still puzzling to the Church authorities. Having heavily promoted the idea that the heavens were built around a sequence of celestial spheres centred on Earth, churches of all denominations were now unwilling to admit to being wrong. To do so could risk ceding authority to the astronomers. As a result, astronomers suddenly found themselves under the same scrutiny – even suspicion – as the astrologers.

In an attempt to ease its reception in religious circles, Copernicus's book contained an anonymous introduction that had been inserted before printing without the author's knowledge. It was almost certainly an attempt to mollify any potential controversy over the proposed rearrangement of the planets.

The person thought responsible for the foreword was Protestant theologian Andreas Osiander, who had overseen the final stages of printing. He wrote that *De revolutionibus* contained ideas that were to be taken as mathematical recipes only; they could be used to furnish better predictions of celestial movements but should never be taken as meaning that the Earth actually moved. An astronomer would always choose the easiest hypothesis, he continued, whereas a philosopher would plump for the one he thinks is right, but neither will know anything for certain because truth can only be revealed through divine revelation. With these words, he sought to reassure the theologians that they retained the ultimate authority, but astronomers were aggrieved by the suggestion that their mathematics was a trick.

Over the centuries the skywatchers had developed a suite of instruments that could transform the night sky into numbers, giving them the raw data they needed to make their calculations. It was a way of intellectually connecting to the night sky and deducing what was happening in a realm that no one imagined they would ever be able to visit. By Copernicus's time the available suite included the cross-staff, astrolabe, armillary sphere and quadrant. Though varying in complexity and accuracy, all served the same basic purpose of measuring the position of celestial objects in relation to a fixed position on Earth.

These measurements could then be checked against the predictions of the various 'models' of the cosmos – such as those of Ptolemy and Copernicus – to see which reproduced the measured observations most closely. But a model is only ever as good as the observations it seeks to reproduce, and in the time before telescopes, the instruments were difficult to use with any real accuracy.

Theologians and philosophers routinely criticised astronomers over their inaccuracy. They pointed out how difficult astronomers found it to measure the details of the planetary motions, and without these, they said, it was difficult to distinguish between the various models of how the cosmos was arranged. Hence, as a matter of expediency, theologians and philosophers argued that the engrained model should simply be retained. Copernicus, however, was cleverer than this. He pointed out an over-looked aspect of planetary motion that clearly discredited Ptolemy's geocentric model.

He looked at the backtracking of the outer planets, and noticed a new detail that Ptolemy's model could not explain: the backtracking of Mars, Jupiter and Saturn only ever takes place when the planets are nearing their highest altitudes at midnight. At no other time did this looping motion set in. Copernicus realised that if it were caused because the Earth is on an inside track and lapping the outer planet, then at the moment Earth pulls ahead, our world will always be directly between the Sun and the planet in question. This will mean that the distant planet will always be directly opposite the Sun in the sky, which will make it appear at its highest altitude at midnight.

With this realisation, Copernicus understood that the Earth had to be moving. Yet try as he might, he still could not make his model fully reproduce the motions of the planets. With hindsight we can pinpoint the reason; it was because he retained the idea of the celestial orbs, which meant that he also retained the idea that the planets moved in perfect circles. This was an error Copernicus remained blind to, ultimately having to resort to introducing epicycles of his own that were even more complicated than the Ptolemaic model he was trying to simplify. Nevertheless, he had established that the Earth was indeed moving through space.

Sales of his book, *De revolutionibus*, numbered in the hundreds rather than the thousands, which led some historians to believe that the book was a failure. The truth, however, is more subtle. The twentieth-century astronomer and historian Owen Gingerich spent thirty-five years tracking down almost three hundred surviving copies of the first and second editions of *De revolutionibus* to look for any notes their readers had made in the margins. In doing so, he found that the book, which was highly technical in presentation, was read by all of the major astronomers of the time, and the notes they made in their copies show that they took the contents seriously.[52] One of those readers was Johannes Kepler, born in 1571 in Weil der Stadt, Swabia (now Germany).

Kepler's interest in the night sky was kindled when he was six and his mother took him outside to show him a great comet stretching across the sky. He encountered the Copernican model of the planets – and the ideas of astrology – while at the University of Tübingen. He became skilled at constructing horoscopes by practising on his fellow

The great conjunctions of Jupiter and Saturn from 1583 to 1723, as drawn by Johannes Kepler (1571–1630). The diagram shows the zodiacal constellations in which these alignments took place. (Wikimedia)

students, but it was his theological education that arguably had the greatest impact on him because he ended up wanting to become a Lutheran Protestant minister.

He was persuaded instead to take a teaching position in Graz, and once there he had a moment of inspiration that would lead to his discovery of the true shape of planetary orbits. He was contemplating one of the most beautiful sights in the night sky, the so-called great conjunction between Jupiter and Saturn. These are quite rare events that only take place every few decades as Jupiter laps Saturn and the two bright planets appear to draw close in the night sky. Jupiter shines brilliantly white, whereas Saturn displays its subtler shade of ochre. Together, the great conjunctions were thought to be astrologically significant because of the pattern they trace in the sky.

The pattern comes about by pure chance and derives from the fact that each successive great conjunction takes place roughly eighteen to twenty years apart. As Saturn takes approximately thirty years to circle the Sun, this means that the ringed planet travels roughly two-thirds of the way through its orbit, or about 240° further across the sky, between successive great conjunctions. Over the course of about fifty to sixty years, three great conjunctions take place defining an equilateral triangle in the sky. The astrologers called this triangle a trigon, and since each point on the triangle falls into a different sign of the zodiac, they thought nature was telling them that these particular constellations were linked in some way.

A further subtlety is that Saturn does not quite complete two orbits during each trigon. This means it does not quite return to its starting position, and that makes each successive trigon shift about 7–8° along the zodiac. Since the astrologers had divided the zodiac into the twelve signs, each of which was 30° wide, four successive trigons would fall into the same set of constellations before moving on and forming another set. As there are twelve zodiacal constellations, this means that there are four sets of three constellations. Each trigon was associated with one of the four classical elements. The earth trigon is Aries, Leo and

Sagittarius; the water trigon is Gemini, Libra and Aquarius; the air trigon is Taurus, Virgo and Capricorn; and the fire trigon is Cancer, Scorpio and Pisces.

To the astrologers, the neatness of it all suggested profound meaning. Comments on the supposed importance of these conjunctions is first definitively found in the writings of Muslim astrologers in Baghdad from the eighth and ninth centuries, although the ideas they write about may have come from Iran a few centuries earlier.[53] Special significance was attached to the years in which the trigon passed from one set of constellations to the next. This occurred after every fourth trigon, or roughly every two centuries, and it was thought to be a moment of historic change. However, the greatest significance was reserved for the time when the whole cycle had passed and was about to start again. This took place roughly every eight hundred years or so and, according to the astrologers, separated history into 'great ages'.

Christian astrologers and theologians reckoned that there had been six great ages since the biblical Creation, and that they represented the times of Enoch, Noah, Moses, the ten tribes of Israel, the Roman Empire and the birth of Christ, and finally the establishment of the Holy Roman Empire by Charlemagne. In 1583, as Kepler was approaching his teenage years, another cycle began and was hailed as a momentous event.[54]

Marking only the second great age since the birth of Christ, the conjunction in 1583 sparked a cavalcade of publications predicting various dooms for European society. A persistent line of thought claimed that the biblical final judgement was coming. To stem the rising tide of public concern, the Pope issued a bull banning all divinatory practices in 1586, but the predictions continued, especially in Protestant countries. In the 1590s, when Shakespeare wrote *Henry IV*, he was either lampooning or simply recalling the furore in Part II, Act II, Scene IV, where he had the characters of Prince and Poins discussing the meaning of a conjunction between Saturn and Venus in the 'fiery Trigon'.

Kepler's interest was centred on the fact that the orbits of Jupiter and Saturn seemed designed to make the great conjunctions take place. He imagined God placing them precisely in these courses during the Creation. By the 1590s, he also believed that God had given him the intuition to perceive the underlying reason for the planets' positions. It rested on invisible geometrical shapes working like scaffolding to hold the planets apart.

In proposing this idea, Kepler had drawn inspiration from the work of Plato. In *Timaeus*, Plato had described the five 'perfect solids'. These are three-dimensional forms that are constructed by fitting together two-dimensional equilaterals. In this system, six squares can be fitted together to make a cube, four equilateral triangles can be fitted together to make a pyramid-like tetrahedron,[55] eight equilateral triangles can be fitted together to make an octahedron, twelve pentagons form a dodecahedron, and twenty equilateral triangles make an icosahedron. These are the only five perfect solids that are possible, and Plato proposed that the first four were the microscopic forms taken by the classical elements, earth, water, air and fire. The fifth shape, the icosahedron, was the celestial aether/quintessence.

Kepler thought much larger versions of the Platonic solids held the planets' celestial orbs apart. He calculated that the spheres of Saturn and the stars could be kept apart by a cube, the spheres of Jupiter and Saturn by a tetrahedron and so on. To prove his grand design, which followed Copernicus by placing the Sun at the centre of the planets, he raised money from Friedrich I of Württemberg to build an actual model of his ideas in silver. It was to stand in the entrance hall of the Duke's palace, where it would be a talking point. Kepler even promised that the pipework would be hollow so that it could function as a drinks dispenser. Each shape would carry a beverage that mirrored the astrological properties of the planet it supported. In the case of Saturn, this meant filling it with a bad beer or corked wine. The Duke could offer his guests a welcome drink from the sculpture, and laugh with derision at anyone uneducated enough to choose to drink from Saturn's cup.

The pieces for the model were manufactured to Kepler's precise specification and brought to the palace for construction – but disaster struck. It didn't fit together. His mistake was that he had followed Copernicus's assumption that the planet's orbits were circular. Humiliated, Kepler sold the silver for scrap and went back to the drawing board, determined to understand what had gone wrong. His embarrassment drove him ever harder towards an eventual breakthrough that would help spark the scientific revolution.

★　★　★

Kepler realised that he had to investigate not just the size of the planets' orbits but their true shapes as well. To do this, however, he needed something he didn't have: a jealously guarded collection of astronomical observations unique in human history for their quantity and precision. They belonged to Danish nobleman Tycho Brahe and were nothing less than his life's work. But they were also quite useless in their current form. They languished in a collection of ledgers as long lists of angles between celestial objects. Only after painstaking analysis using an astronomical model would they reveal the size and shape of each planet's orbit.

Now in his advancing years, Tycho was desperate to know these orbits so that he could predict the planets' future positions, thus leaving a precise celestial almanac as his legacy to humankind. Yet he did not possess sufficient mathematical flair to dissect his observations. On the face of it, Kepler and Tycho were perfectly matched. Kepler had the brainpower, Tycho had the data. But when the two met on 4 February 1600 at Benátky Castle, near Prague, problems soon erupted.

Tycho clung to the old geocentric model, whereas Kepler was certain that Copernicus was correct. They fell into bitter argument. Each needed the other, yet neither was able to accept their differences. In the end, tragedy intervened.

Tycho died suddenly, probably of a bladder infection, and in the resulting confusion Kepler purloined the observations and fled. He then wrestled with the data for decades, filling thousands of pages with calculations, fighting what he later called his 'war with Mars' until he found the mathematical shape that perfectly reproduced Mars's motion through the night sky: it was an ellipse, not a circle. The key to finding it had been to first remove the effects of Earth's own orbit. Once he did this, all the retrograde loops disappeared, leaving the true movement of Mars. Emboldened by this success, he analysed the observations of the other planets and found that they too follow their own ellipses. In the end, he found he could summarise the way all the planets move in just three lines of mathematics. Known as Kepler's laws of planetary motion, they are still one of the first things taught on any astronomy degree.

The first of Kepler's laws is that planets move in elliptical orbits around the Sun. The second is a mathematical description of how a planet speeds up and slows down depending on where it is in its orbit. The third law relates the average speed of a planet to the size of its orbit, showing that more distant planets move with slower average speeds.

It is no exaggeration to say that these simple laws are a watershed in history. Kepler's laws hold true for every planet in orbit around the Sun, including the ones that Kepler didn't even know existed and were only discovered centuries later. They also explain the movement of the thousands of other planets that, in the last few decades, our modern telescopes have discovered in orbit around other stars.

Beyond the science that these achievements would spark, the cultural value was incalculable. In the act of measuring the positions of the stars, Tycho Brahe had captured nature and transformed it into numbers. Then Kepler had used mathematics to distil that information into something meaningful: a precise model of planetary motion. It was a staggering call to intellectual arms that proved it was not scripture, but measurement and intellect that could unlock the secrets of the night sky.

Kepler published a grand synthesis of his ideas in 1619, in a book entitled *The Harmony of the World*. And in that book, he transformed the planets' various elliptical orbits into musical scales.

According to Kepler, the ellipticity of a planet's orbit determined the range of notes in its scale. For example, the most elliptical path was followed by Mercury, and so it swept through the greatest range of musical notes. Venus's orbit, on the other hand, was almost indistinguishable from a perfect circle. If it had a range of notes at all, said Kepler, they must be separated by a diesis – a small, highly dissonant interval.

Earth, he said, ranged between two notes a semitone apart. He even labelled these as the notes Mi and Fa from the eleventh-century Solfège system of note names. Ever despairing of life's difficulties, Kepler noted that Mi and Fa were the perfect notes for Earth, where misery and famine held sway.

As well as defining the note ranges that the planets would sing, he also assigned them to voice ranges. Mercury was a soprano, Earth and Venus were the altos, Mars the tenor, Jupiter and Saturn the basses.

Like Gaffurio, he thought that as the planets moved, so a complex polyphony was produced. It was a kind of motet – a composition of interweaving voices that created an ever-changing harmony. He realised that as these six lines wove together they would mostly produce dissonances, but he felt hopeful that there would also be passing cadences and moments of beauty.

He also realised there would be something else slightly alien about the cosmic music. The notes from each planet would not sound in discrete intervals, moving from one note to the next. Instead, they would slide one into the other like a siren. This was because the planets move smoothly through their orbits rather than jumping from one step to the next.

Another question Kepler tackled was whether the music ever repeated. To do that, the planets would have to return periodically to their starting configurations, and Kepler realised that it was practically impossible to synchronise six planets to do this. He therefore suggested that during God's Creation of the universe, the planets had been placed to produce a perfect cadence, but that glorious harmony would never be repeated.

Kepler's interpretation of the planetary orbits in musical terms is without doubt one of the greatest confections of the human mind, and as others became familiar with his work, they questioned the need for such intellectual decoration. Kepler's laws can be used purely mathematically to predict the locations of the planets in the future, and their positions in the past. This was the essential information that was needed to make sense of the night sky, and it relied only on mathematics, not music.

With this realisation, the idea that music rules the universe collapsed. The natural philosophers who followed Kepler became ever more indistinguishable from today's scientists, in that they started looking for mathematical relationships rather than musical ones to understand astronomy and the rest of nature. In 1623, the Italian astronomer

Galileo Galilei – who we will soon meet – expressed this opinion in his book *The Assayer*.

> *The universe cannot be read until we have learned the language and become familiar with the characters in which it is written. It is written in mathematical language, and the letters are triangles, circles and other geometrical figures, without which means it is humanly impossible to comprehend a single word. Without these, one is wandering about in a dark labyrinth.*

And Kepler wasn't finished yet with remaking our understanding of the night sky. The next question he tackled was *why* the planets moved. Kepler knew that the idea of celestial orbs carrying the planets was no longer tenable because of observations that Tycho had made in 1577. Back then, the Dane had observed the same comet that had sparked Kepler's interest in the night sky. Whereas Kepler had still been a child, Tycho commanded an ensemble of astronomers on the Danish island of Hven and was in the process of building an extraordinary observatory for them to use. Named Uraniborg in tribute to Urania, the Greek muse of astronomy, it featured large floor-mounted versions of the usually hand-held observing instruments. Among the equipment was an armillary sphere 1.6 metres in diameter, a quadrant two metres across, and a stripped-down armillary sphere that spanned a giant three metres. By oversizing them, the accuracy with which he could read the angles increased because the scales were so much larger.[56] Even though they were working before the invention of the telescope, his skilled astronomers read angles

The Ishango bone was found in 1960 in the Democratic Republic of the Congo. It displays a set of marks that archeologists have interpreted in many ways, including as a possible lunar calendar. (agefotostock/Alamy)

This painting from the tomb of Ramses VI depicts the Egyptian sky goddess Nut. Every evening, she swallowed the Sun, which travelled through her body at night and was reborn at dawn. (Hans Bernhard/Wikimedia)

The Dendera sky map was found in the ceiling of the Temple of Hathor, Dendera, Egypt. It is a bas-relief showing a detailed pictorial representation of the night sky. It now resides in the Louvre, Paris. (Alamy)

A modern telescopic image of the Pleiades star cluster. Also known as 'the seven sisters', they are the only obvious tight-knit collection of stars in the whole sky. (Nik Szymanek)

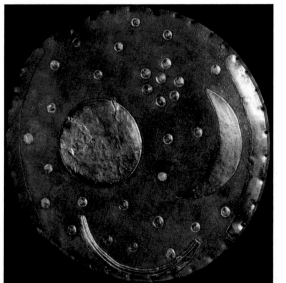

With a diameter of around 30 centimetres and a weight of 2.2 kilograms, the Nebra Sky disc is generally thought to be a representation of the night sky. Dated to around 1600 BCE, it shows a full moon (or the Sun), a crescent moon, stars and the Pleiades star cluster. (Heritage Images/Getty)

The Hall of the Bulls in the Lascaux caves, France, has attracted much attention from astronomers because the six spots at the top right of the image mirror the location of the Pleiades star cluster with respect to the constellation Taurus, the Bull. It is also suggested that the four dots the Bull is facing could be a representation of the stars in Orion's belt. (Patrick Aventurier/Getty)

The Farnese Atlas, on display in the Museo Archeologico Nazionale, Naples, Italy, dates to the second century. The sky is blank where Atlas bears the globe on his shoulders, indicating that the inventors of the western constellations were located in the Mediterranean regions and so could not see these extreme southern parts of the sky. (Adam Eastland/Alamy)

A miniature from *The History of Mohammed* depicting the paradise that awaits in the 'seventh heaven', which was thought to be the realm of existence beyond the visible celestial spheres. (Granger Historical/Alamy)

Uon dem donnerstein gefallē im xcij .iar:vor Enſiſheim:

Ensiſheim

Battenhem

This woodcut shows the fall of the Ensisheim meteorite in what is now Alsace, France. Because of the testimony of eyewitnesses, this is the oldest meteorite whose fall can be dated precisely, to a few minutes before noon on 7 November 1492. (Detlev van Ravenswaay/Science Photo Library)

Joseph Wright of Derby (1734–97) painted *A Philosopher giving That Lecture on the Orrery, in which a Lamp is Placed in the Position of the Sun*, to illustrate the rise of popular scientific understanding in the eighteenth century. (Derby Museums)

Adam Walker (1731–1821) and his sons enthralled audiences with their spectacular machine, the Eidouranion, and their poetic descriptions of the cosmos. (V&A)

Vincent Van Gogh (1853–90) was inspired to paint his masterpiece *The Starry Night* during a spell in the asylum of St Paul de Mausole, Provence, France. (Peter Horree/Alamy)

The Milky Way stretches high into the sky above the Anglo-Australian Observatory, Siding Spring, Australia. It was here as a research student that I experienced Immanuel Kant's notion of the cosmic sublime first hand. (Science Photo Library)

Earthrise is one of the most extraordinary images to come out of the mid-twentieth century space race. The crew of Apollo 8 were in orbit around the Moon during December 1968 when they spotted and photographed our world rising over the lunar horizon. (NASA)

Astronomers gather at twilight on the summit of Cerro Paranal, the site of the European Southern Observatory's Very Large Telescope (VLT) to greet the night. It was while standing in that same location in 2005, that I realised what it truly feels like to be a human individual surrounded by a vast, beautiful universe. (ESO/G. Brammer)

Chesley Bonestell (1888–1986) captured the public's imagination with his 'photo-realistic' views of our Solar System. Here he shows what Saturn would look like from the surface of its moon Mimas. He often included tiny astronauts for scale. (Chesley Bonestell)

Perhaps the quintessential image of the space race is Astronaut Edwin 'Buzz' Aldrin walking on the surface of the Moon. It was taken on 20 July 1969, when Aldrin and mission commander Neil Armstrong became the first humans to walk on another world. (NASA)

that were just one hundredth of a degree across. They measured the length of the year to within a second of its modern value, and the tilt of the Earth's axis to within a hundredth of a degree.

At this time in history, most thought that comets were atmospheric effects. The German painter and astronomer Georg Busch had even published the opinion that comets 'are formed by the ascending from the earth of human sins and wickedness, formed into a kind of gas, and ignited by the anger of God. This poisonous stuff falls down again on people's heads, and causes all kinds of mischief, such as pestilence, sudden death, bad weather and Frenchmen.'

Tycho's observations of the comet had been taken with a 65-centimetre brass quadrant, and when Kepler analysed them he saw that the celestial object was crossing the orbits of the other planets unhindered. Hence, there could not be solid celestial spheres carrying the planets as Ptolemy had hypothesised and the world had accepted. And Kepler didn't stop there; he and Tycho dealt a fatal blow to another central pillar of astronomical belief: that the heavens were perfect and unchanging. This idea had been established by Aristotle and suggested that God – the perfect being – would be incapable of creating the heavens in anything other than perfect form. This was the explanation given for why the stars appeared unchanging in their constellations night after night, year after year.

Yet, both Kepler and Tycho saw new stars appearing in the heavens. These 'novae' as they became known were shown in the twentieth century to be erupting or even exploding stars that become temporarily bright enough to be seen by the naked eye. But to Tycho and Kepler, to kings and leaders of the time, and to the general population as well, the sudden appearance of new stars was extraordinary. Their discovery was a hammerblow to the Aristotelian idea of the divine and unchanging heavens, because any perceived change would imply a move away from perfection, or that they hadn't been perfect in the first place. When Tycho saw his nova in 1572 and Kepler saw his in 1604, both struggled to explain the consequences. Eventually, they both settled on their events having astrological and religious meaning.

Tycho associated his nova with the great conjunction in 1583, and warned of consequences that would last decades, although he made no specific predictions. Kepler's nova of 1604 came at a particularly thorny time and place in the night sky because it coincided with a great conjunction of Jupiter and Saturn. This was the first great conjunction to take place in the fire constellations and supposedly marked the beginning of a new great age. By coincidence, Earth had lapped Jupiter at the time, causing the planet to backtrack in the sky, and this had caused a triple conjunction whereby Jupiter had appeared to circle back to pass Saturn a second time, and then a third time as Jupiter continued its forward motion again. And then, just when the drama appeared to be over, Kepler's nova appeared near the location of the conjunction.[57]

Kepler decided that it was all of great religious significance and that it must have been created as a message from God to Earth. Doubtless thinking about the prophesied last judgement, he urged readers of his book *De Stella Nova* to repent their sins. Of course, no such great cataclysm came to pass, but by the time Kepler published his first two planetary laws in 1609, an earth-shaking event of a different kind – one that would forever change the way we think about the universe – was on the horizon.

★   ★   ★

Galileo Galilei was a lecturer in mathematics, geometry and astronomy at Padua University in Italy. In 1609, he heard of a device invented by the Dutch spectacle maker Hans Lippershey that could magnify distant objects. The rumour reached him around the middle of May, and he immediately set about grinding lenses to see if he could reproduce such a device.

Galileo toiled throughout summer and autumn and, when the sun set on 30 November, he lifted up his invention – the telescope – and pointed it at the night sky. He targeted the Moon, a waxing crescent just four days old, and noted how the sunlight was creeping across the lunar surface. He charted it for the next five days, discovering that the Moon was both mountainous and covered with craters.[58] On 18 December, he aimed the telescope at the Milky Way and made his next revelation when he saw the misty band of light resolve into individual stars. On 7 January 1610, he observed Jupiter and discovered three 'stars' close to it. The next night, he noticed that the position of the three stars had changed their configuration. He realised at once that he was witnessing something unprecedented, but clouds frustrated his 'strong desire' to re-observe the planet the following night. He continued watching and soon discovered a fourth 'star' near Jupiter. Just five days later, he realised

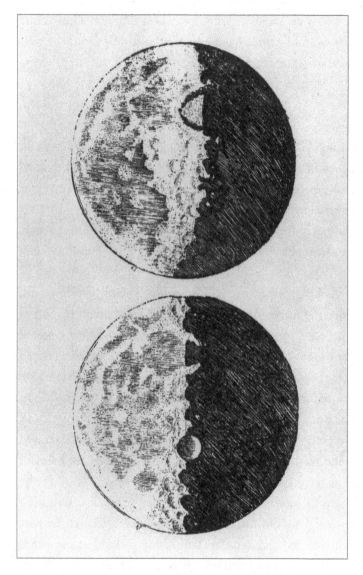

Using a telescope, Galileo discovered that the Moon was a mountainous world similar to Earth. This contradicted the accepted idea that it was perfect sphere made of a celestial substance called aether. (Science & Society Picture Libary/Getty)

that the four stars were in orbit around the planet – they weren't stars at all, they were moons.

Realising the utterly historic nature of his discoveries, the very next day he began preparing his observations for publication, and by 30 January he was in Venice liaising with the printer. By 19 March he was sending out finished copies of the book, *Sidereus Nuncias* (The Starry Messenger).[59]

Galileo's observations were a total dismantling of Aristotle's ideas. The craters and mountains of the Moon proved that it was not a perfect sphere. Instead the landforms made it look distinctly Earth-like – no aether in sight. The moons of Jupiter introduced their own problem, because it showed that there was a centre of revolution in the cosmos that wasn't around the Earth. And the more Galileo looked, the more questions he found. At the end of 1610 he saw that Venus displayed phases like the Moon; direct evidence that it followed an orbit around the Sun. Although some claimed that this only proved heliocentricity for Venus (and by extension Mercury), the truth was that taken together with Kepler's laws, the old geocentric cosmos was untenable. The observations made it clear: the Earth orbited the Sun, just like the other planets.

Famously, Galileo was tried by the Roman Catholic Inquisition for his insistence on the demonstrable truth of his observations. The first run-in took place in 1616, when he was warned to refer to his ideas as hypotheses only. Galileo crossed that line time and again by asserting that the Earth moved in reality, but his greater crime was to suggest reinterpreting the Bible to accommodate this view. Biblical interpretation was the sole province of the Vatican theologians. It was the bedrock on which the authority of the Catholic Church was based. Yet Galileo continued to pursue the new astronomy vociferously, partly because he feared that the Protestants would end up overtaking the Catholics in terms of progressive knowledge, and partly because he thought of his work as discovering a fundamental truth about God's realm, and therefore something that was religiously justified.

The Vatican's patience with Galileo ran out in 1633. He was tried and found to be 'vehemently suspected of heresy'. He then recanted his astronomical opinions, and was held under house arrest for the rest of his life. This moment in history is often held up as a case of 'religious dogma versus scientific truth', particularly after the twentieth-century German dramatist Bertolt Brecht used it as an excoriation of authoritarian rule in his play *The Life of Galileo*. In truth, it was more complicated than that.

In the 1620s, the Jesuits of the Roman College were coming round to Galileo's way of thinking. Moreover, they were beginning a careful discussion with the Vatican theologians about changing the interpretation of the Bible. But the stumbling point was that no observation categorically proved the Earth moved. Indeed, the best test, which looked for a phenomenon known as parallax, was most easily interpreted as showing that the Earth was stationary.

Parallax is the change in position of a foreground object against the background when viewed from a different perspective. It is easily demonstrated by holding up your index finger, about a foot in front of your face. Close one eye and notice where your finger falls in relation to the background. Now swap eyes but don't move your finger. You will immediately see that despite remaining stationary, your finger appears to have jumped across the backdrop. Repeat the experiment with your finger at arm's length. This time, the distance your finger appears to move is much less. Thus, if the Earth were moving, nearby stars should move in relation to more distant ones. But they didn't seem to. No parallax was observed.

We now know this was because the stars are so far away that the telescopes of the time were too weak to show the phenomenon. Indeed, technology did not improve sufficiently until the nineteenth century. But the seventeenth century's failure to detect it meant that theologians were not willing to change their interpretations of the Bible: the success of Kepler's laws, which assumed a moving Earth in their prediction of planetary motion across the sky, was not seen as strong enough evidence without the observation of parallax to back it up.

In hindsight, it is easy to brand this a stupid, dogmatic decision, but at the time it was more a practical one. If a thousand years of scriptural belief and cultural wisdom were to be overturned, then those in power had to be in possession of overwhelming evidence to justify such an upheaval. And at the time, the Church was much more concerned about some of Galileo's other observations, because these had been verified by the Jesuits, and were a cause of grave theological concerns.

The observations in question were the ones that showed myriad stars across the night sky that were invisible to the naked eye. As soon as Galileo raised his telescope, he could see so much more than before. Now we take this completely for granted, but at the time it was a complete shock, and raised a question that struck at the very heart of their understanding of the night sky and our place relative to it. Why would God hide things from our view?

In pursuing the answer, the West continued to redefine our relationship with the night sky, pushing it ever further from the old certainties and eventually cutting them adrift entirely. According to some psychologists, this shift in our understanding of the night sky's role in our lives was so profound that we are still suffering from the shock.

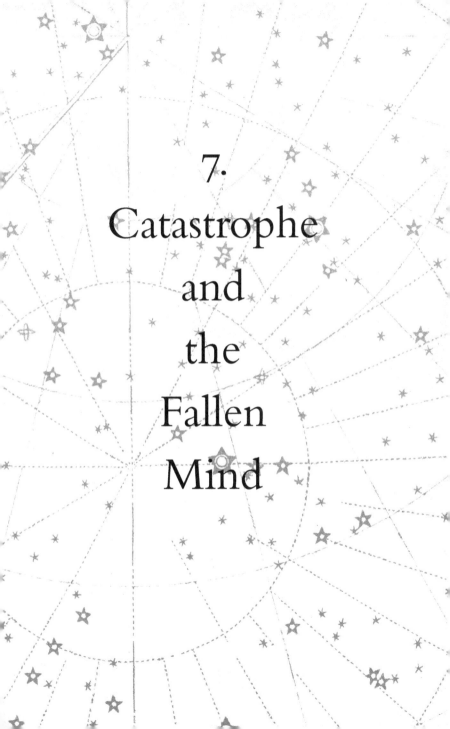

7.
Catastrophe
and
the
Fallen
Mind

The Christian philosophy of the world was that God had designed everything to be of use to humankind. When it came to the night sky, the predominant view was that it was a time-keeping mechanism. The planets and other celestial objects moved across the clock face of the universe, on which the constellations were used to mark their progress. So when Galileo and others found so many otherwise invisible stars strewn across the night sky, it threw that view into utter confusion. The dilemma of why God would have hidden these stars from us fed a theological debate that had been gathering momentum for several centuries about the concept of original sin and the Fall of Adam and Eve after eating from the forbidden tree in the Garden of Eden.

The debate centred on the precise nature of the Fall and the impact it had on us. In a 1662 address to the congregation in St Paul's Cathedral, London, the preacher Robert South described Adam as coming into the world as a philosopher who had God-given perfect knowledge of nature and the cosmos.[60] As part of God's punishment for eating the forbidden fruit, that understanding of the cosmos was taken from Adam. But exactly how much knowledge had been taken from us?

In a work published in 1620 called *The Great Instauration*[61] the English philosopher Francis Bacon described how the emerging 'mechanical philosophy' of people like Kepler and Galileo could rediscover Adam's lost knowledge and answer the question about how much knowledge had been lost in the Fall.

Mechanical philosophy was so named because it assumed the cosmos worked like a giant mechanism, with strict rules of cause and effect. Bacon speculated that if we based our theories only on observation and experimentation, our understanding of nature and the surrounding cosmos could be fully returned to its original, perfect condition. It was a heady concept, which he detailed in a follow-up book, *Novum Organum*, also published in 1620. He warned against basing new ideas on older untested ones, or extending ideas beyond what could be demonstrated experimentally. This was the beginning of the scientific method, which stresses the need to continually test our ideas with further investigations.

Robert South could well have read Bacon's work. The timing of his address came almost two years to the day after King Charles II granted a royal charter to inaugurate the Royal Society of London for Improving Natural Knowledge. This group of individuals were dedicated to the Baconian investigation of nature. They even adopted the motto *nullius in verba*, meaning take nobody's word for it.

Galileo's telescopic discoveries suggested a new way to interpret the Fall. Instead of just tearing the knowledge from Adam's mind, perhaps God had altered the cosmos to obscure most of it from our view, or simply dulled our senses. In either case, we lost our ability to see all the stars. This meant that far from being a threat to religious authority, the invention of the telescope was actually pivotal to humanity's salvation – it was the instrument for the rediscovery of Adam's lost knowledge and the atonement for original sin. At the other end of the size scale, the newly invented microscope was performing the same function, showing us things too small to be seen with the naked eye. The great English experimentalist Robert Hooke wrote as much in the preface of *Micrographia*, his 1665 book that presented his drawings of the invisible world seen through his microscope.

Adherence to the new mechanical and experimental philosophy brought many old ideas in for fresh scrutiny, and this led to the final abandonment of Aristotle's theories of physics as well as the classical elements of earth, water, air and fire. In 1611, as the potential ramifications of the new astronomical observations started to become clear, John Donne lamented in his poem *An Anatomy of the World*:

And new philosophy calls all in doubt,
The element of fire is quite put out,
The sun is lost, and th'earth, and no man's wit
Can well direct him where to look for it.

This uncertainty persisted for most of the 1600s. It was only in the later half of the century that the blueprint for a new certainty presented itself. When it did arrive, it came from an unexpected quarter: an eccentric Cambridge recluse by the name of Isaac Newton.

★　★　★

It began in January 1684 when astronomer Edmond Halley met with Christopher Wren and Robert Hooke. All three were mechanical philosophers, members of the Royal Society and in pursuit of the underlying factor that forced the planets to obey Kepler's laws. They could see that those laws implied some kind of influence was emanating from the Sun, but the strict mathematical description of its behaviour eluded them.

Of the three men, Halley was the best mathematician, but even he was stumped. He could think of only one recourse: to approach Isaac Newton, an awkward yet gifted natural philosopher who had shunned the Royal Society more than a decade before to live as a virtual recluse in Trinity College, Cambridge. Wren encouraged Halley to make the trip. Hooke was against, having clashed with Newton over a discussion about the nature of colours in the 1670s.

Ignoring Hooke, Halley travelled to Cambridge in August 1684, and asked for help. To his amazement, Newton claimed to have already performed the necessary calculations, but when Halley asked to see the proof, Newton conducted a cursory search and proclaimed the paper lost. He promised to redo the maths and send it down to London.

Months passed, and then at a meeting of the Royal Society in November a small document was handed to Halley. It was the proof he had hoped for – and much, much more. Newton had clearly been gripped by the problem and had succeeded in deriving all three of Kepler's laws using the assumption that a force was coming from the Sun. He called this force 'gravity' and showed that to make the planets move as Kepler described it had to behave in a very specific way. If one planet were twice the distance of another, then the gravity acting on the further one would be a quarter of that acting on the nearer. If it were triple the distance, the force would be one ninth. This is known as an inverse square law and crops up a lot in physics. For example, it also governs the way a source of light dims with distance. So, if one star were located twice as far away as its identical twin, it would appear one quarter as bright.

Newton had gone on to generalise Kepler's first law to show that the precise shape of an orbit was determined by the average velocity of the celestial object. This meant that ellipses were just one possible shape. This fact would become critically important to Halley when he began considering the shape of comet orbits. But for the moment, stunned by what he read, Halley raced back to Cambridge to seek Newton's permission to circulate the document. Newton refused.

He told Halley that he had glimpsed a way to extend the work beyond the planets to encompass the movement of all things. If he was right, he was on the cusp of an entirely new science of how forces made all things move. The key to it was gravity, working at all scales across the cosmos and down here on Earth as well. He thought that an object fell to the floor for the same underlying reason that the Moon circled the Earth. Like the first time Halley came to see him, Newton didn't yet have the mathematical proof. But this time, both men realised just how much was at stake.

For thousands of years – if not tens of thousands – humankind's view of the night sky was based on the assumption that it was a fundamentally different realm from the Earth, and therefore subject to different rules. If successful, Newton would upend this notion entirely. A unified mathematical description of celestial and earthly motions would prove that exactly the same natural laws operated in both places.

Newton became so consumed by his computations that he ate only sparingly, often forgetting to eat at all. He would absent-mindedly go about his daily tasks before shrieking with inspiration and rushing to his desk, where he would often work standing up, having forgotten to pull up a chair. After three years of sustained effort, he finished the work. The result was the greatest scientific masterpiece in history: *Philosophiæ Naturalis Principia Mathematica* (Mathematical Principles of Natural Philosophy). Published in July 1687, it was a labyrinthine work of mathematics that showed how motion was always the product of a force. If something moved, the underlying force propelling it could always be calculated. Sometimes the force was obvious, such as a horse pulling a cart or a person throwing a ball into the air. Other times, the force was invisible, such as when a pebble rolls downhill or a planet crosses the night sky. Newton showed that the invisible force was always gravity, an attractive force generated by all objects with mass.

As astonishing as it may seem, rain falls to Earth for the same reason that a shooting star burns across the night sky: because gravity is acting. The only factors that affect the strength of gravity are the masses of the two objects involved and the distance between them. In showing this, Newton unified Kepler's work on celestial motion and Galileo's work on earthly motion. Moreover, Newton's mathematical formula for gravity allowed predictions to be made and tested with observations. If the predictions were shown to be true, then regardless of whether anyone liked the idea or not, it had to be accepted as the truth – and this was where Halley and his comet came in.

As well as providing the initial spark for the work, the astronomer had also shepherded *Principia* into print. He had acted as editor, publisher and even bank-roller for the project when The Royal Society exhausted its finances publishing *The History of Fishes*, which flopped.[62] Now, he turned to using gravity as a means of understanding comet orbits.

We now know that comets are mountain-sized chucks of ice and rock that condensed far away from the Sun when the planets were forming, but in Halley's time they were still somewhat mysterious. Tycho and Kepler had shown that they were celestial objects but they clearly followed very differently shaped paths than the planets. To prove this in *Principia*, Newton had analysed two comets, one of which had appeared in the evening skies of 1680 and one from the morning skies of 1681. He reasoned that they were the same object travelling on a highly elongated orbit around the Sun, and with this assumption he found that the two trajectories could be understood as different lengths of a tightly curving path known as a parabola.

From here, Halley took up the challenge and set about finding the orbits of twenty-three other comets that had appeared between the years 1337 and 1698. As he worked on the data, three comets stood out immediately. They appeared in 1531, 1607 and 1682. All followed similar paths to one another. Upon further investigation, Halley proposed that these were actually the same object returning again and again. And if so, based on the elliptical orbit he calculated, it would next be seen in late 1758 / early 1759.

On Christmas Day 1758, German farmer and amateur astronomer Johann Georg Palitzsch saw a comet appear in the sky. A quick check of its location proved that it was indeed Halley's comet, returned as predicted. This was a defining moment of the scientific revolution. What began when Copernicus championed the moving Earth was now crowned with Newton's unification of earthly and celestial physics. The achievement was proof that mathematical analysis could reveal the cause of physical events, and reveal the order of the cosmos. It demonstrated the power of science to predict the future (thus beating the astrologers at their own game), and it set the stage for an outburst of scientific investigations of everything that continues to this day. This huge cultural upheaval was known as the Enlightenment, in which tradition and hierarchy were re-examined with the newly prized qualities of reason and science.

The significance of this moment in history is impossible to overstate. But there was also a down side. As Edmond Halley and others were starting to realise, not all of their new discoveries were pleasant ones. Indeed, in his consideration of the comets, Halley had found something terrifying.

★ ★ ★

In calculating the highly elliptical paths followed by the comets, Halley saw how these mysterious celestial objects came out of deep space and crossed Earth's orbit to shoot around the Sun before heading back into the darkness, crossing Earth's path once more. Halley realised that in crossing our orbit, a comet could potentially collide with us.

He knew he had to bring his findings to the attention of his peers but he feared that doing so would bring him into conflict with the religious authorities and lay him open to the accusation of scaremongering. So, with a promise that his findings would not be published for a wider audience, he took his place in front of the Fellows of the Royal Society one day in December 1694 and spoke secretly about how the end of civilisation could come from the night sky.

It was a chilling thought, and he began thinking about whether there was any historical evidence for such a catastrophe having already happened. His mind went immediately to the biblical deluge. In the Old Testament's Book of Genesis there is a retelling of a flood myth that originated in Mesopotamia in the epics of Ziusudra and Gilgamesh. In the Christian version, God pours water from the sky for forty days and forty nights to submerge the world and destroy humankind for their sin and wickedness. The only people saved are Noah, his sons, and their respective wives. Noah builds an ark on which he loads a male and female of each animal species, and they float out the divine genocide.

What sparked Halley's interest was the cause of the flood. He looked at the weather records for the rainiest counties in England and calculated that even at a constant rate of forty days and forty nights, such rainfall would not be enough to submerge the whole Earth because it would total just twenty-two fathoms (or about forty metres) of water. Hence, only coastal regions would be affected. To extend the flood inland would require something altogether more catastrophic. Given the day-to-day absence of miracles, he noted that the Almighty generally made use of natural means to bring about his will and so proposed that a comet collision could somehow cause the oceans to engulf vast areas of land, bringing about the flood.[63]

As for the place of the impact, he suggested that it would leave a great depression such as the Caspian Sea or one of the great lakes found across the world. Perhaps his most inventive idea was that the North American continent was particularly cold in winter because it had once been at the north pole. The comet's impact had then knocked it somehow away from the polar region, but vast reserves of ice remained frozen beneath its soil.

The presentation itself generated interest and much discussion among the Fellows, and an emboldened Halley returned a week later with some further thoughts that he revealed under the same conditions as before. He said that a person whose judgement he respected had approached him and suggested that a comet colliding with the Earth could be even more catastrophic than simply causing the biblical flood. In the opinion of this other person, the collision could have happened before the biblical creation, utterly destroying the former world. Out of this chaos, our world was then created by divine intervention. Halley went on to suggest that perhaps at some future time, another similar catastrophe might befall our world if God judged it to be the best way of ensuring its future wellbeing.

Halley's prudent concern that his work remained unpublished in the Society's archives was not one shared by all. Just two years later, in 1696, a former student of Isaac Newton's published essentially the same idea in a book called *A New Theory of the Earth from its Original to the Consummation of*

All Things. The author was William Whiston, a promising Cambridge academic who drifted more into controversy and prophesy as time went by. He was finally expelled from the university for heresy against the Anglican Church, namely a rejection of the Holy Trinity,[64] and so turned to giving public lectures about science. He also suggested a cash prize for the first person to suggest a way of determining longitude at sea, a suggestion that was taken up by Parliament in the Longitude Act of 1714.

In 1736, however, Whiston went too far. He panicked London with claims that an approaching comet would destroy the Earth on 16 October by sparking mass fires or even colliding with our planet. As anxiety rose, the Archbishop of Canterbury stepped in and claimed the prediction was false – which indeed it proved to be. In the wake of the affair, Whiston became a laughing stock to the general public and a social outcast among his peers.

And yet one did not have to dig too far into the historical records to find real evidence of rocks falling from the sky. One particularly celebrated example was the great tale of the 'thunderstone' of Ensisheim, Alsace, now in France.

*   *   *

According to reports from the time,[65] a thunderous explosion echoed across the Upper Rhineland a few minutes before noon on 7 November 1492. It drew people out into the wintry landscape from across the region, which included the walled city of Ensisheim and many mountain villages. All looked for the cause of the racket. A boy who had been standing outside the city already knew the answer. He had seen a huge stone rip through the sky and fall in a field nearby. He led the city folk to the place and they stood in wonder, gazing at a black boulder that sat in a small crater, about one metre deep. Organising themselves into a team, they pulled the triangular-shaped lump onto level ground and began chipping off pieces to carry home for good luck. When the city magistrate turned up, he put a stop to the souvenir hunting and had the space rock hauled to the church inside the city.

King Maximilian rode in about two and a half weeks later, on his way to make war on the French. He and his advisers inspected the rock and decided (perhaps unsurprisingly) that it must be an omen of good fortune. They then struck off two fragments to carry as talismans, and proceeded into battle. A large chunk of the stone remains on display to this day in Ensisheim's Musée de la Régence.

Today we call these falling rocks meteorites. They are pieces of the first planetary bodies to form in our solar system, 4.6 billion years ago, and are mainly composed of stone and metal. They orbit much closer to the Sun than the comets. We now know that the metal meteorites in particular are interesting because they were once buried in the hearts of new-born planets that didn't survive to maturity like the Earth. Instead, they were shattered back into fragments by the kind of giant impacts that Halley and Whiston were speculating about.

Back in the fifteenth century, meteorites were another piece of evidence that the cosmos consisted of the same stuff that made Earth, not some special material with divine qualities as Aristotle had claimed. And Ensisheim wasn't the only example. A small handful of other meteorites had been recorded as falling from the sky. The earliest dates from 861 in Nogata, Japan, where a brilliant flash was seen across the night. The next morning, the locals found a fist-sized meteorite in a small crater. To this day, the space rock is kept in the nearby Suga-Jinja shrine. It has become a symbol of the Shinto religion, one of the surviving forms of animism – the original human religion in which everything was believed to possess a spirit. It is brought out for public view only once every five years, when it is carried through the streets in a decorated cart at the head of a parade.[66]

From earlier in history, many Greek and Roman temples have pieces of meteorite enshrined within them. Although modern analysis sometimes shows that the rocks are not actually from space, the fact that the items were venerated for their supposed extraterrestrial origin is the important aspect. Many myths and folk stories have also been concocted around meteorites over the centuries. These usually centre around seeing the bright streak of a shooting star, which is generated by the meteorite passing through the atmosphere.[67] For example, in Swabia, now in Germany, a shooting star was thought to be a sign that the coming year would be good. In Switzerland, it was thought that a shooting star was a divine message that God was warding off a pestilence. In Chile, a shooting star was a sign of good luck, but to activate that luck you must quickly pick up a stone from the ground. In Japan and Hawaii, the good luck from a shooting star can only enter the body if you loosen your clothing.[68]

Although the sighting of shooting stars has usually been associated with good luck, the work of Halley and Whiston drew attention to the fact that large meteorites could cause major damage, even widespread destruction. If we had any remaining doubt about this, it was erased shortly after dawn on 15 February 2013 when a chunk of rock measuring twenty metres in diameter, and weighing in excess of twelve thousand tonnes, came barrelling into our atmosphere near the Chelyabinsk region of Russia. The fireball it created was brighter than the Sun, and partway through its journey it exploded, producing a shower of meteorites and creating a shock wave that damaged more than seven thousand buildings in the region. Thousands of people were injured, mostly from windows shattered by the passing shock wave. Thankfully, no one died and, as impacts go, Chelyabinsk was pretty small. To see something bigger, we need to look back just over a century ago, to the Tunguska impact.

Tunguska is a remote, almost uninhabited region of Siberia. On 30 June 1908, a huge explosion rocked the location. Computer reconstructions suggest that the Earth was struck by a meteorite – or perhaps more accurately a small rocky asteroid – between 60 to 190 metres across. It released hundreds or even a thousand times the energy of the atomic bomb dropped on Hiroshima at the end of the Second World War, and flattened some eighty million trees. The blast wave stripped the trees of their branches, making them resemble telegraph poles. There were no recorded deaths from the event, but one eyewitness gave a hair-raising account of that unexpected morning. He was a trader called Semenov, who was sitting outside his house at breakfast time.

He saw the sky 'split in two', with fire appearing high over the forest. It rapidly spread across half the sky and he was engulfed by an unbearable sensation of heat, as if his shirt were on fire. Before he could act, the rent in the sky closed and a giant blast of sound knocked him backwards by a few metres. As he was lying insensate on the ground, his wife appeared from inside the house and helped him to shelter. But the ordeal wasn't over yet. A thunderous roar battered their ears, as if 'rocks were falling or cannons were firing'. He lowered his head to the ground expecting that at any moment falling rocks would smash him to death. But the ordeal passed, leaving him alive. Around him the great heat had left scorch marks on the ground and the mighty noise had shattered windows.[69]

As terrifying as that sounds, it is nothing compared to what scientists now believe happened in the Yucatan peninsula of Mexico sixty-five million years ago. A vast crater now buried beneath ocean sediment has been discovered just off the coast of Chicxulub. To create such a structure, an asteroid around ten kilometres in diameter must have collided with our planet. Intriguingly, the timing of the impact is roughly coincident with a mass extinction event called the Cretaceous-Tertiary (K-T) extinction, where three-quarters of the plant and animal species of the Earth died out. Among the dead were the last of the dinosaurs.

Although it has not been proven beyond all doubt, the scenario postulates that the heat from the giant impact sparked fires around the globe. As forests and other plant life burst into spontaneous flames, the debris thrown up in the collision came crashing back to Earth, crushing anything unlucky enough to be in its path. This deadly rain reached around the world, as some of the debris had been blasted up to the very edges of space before it fell back down. A vast quantity of dust and other light material remained suspended in the atmosphere for years afterwards, blocking the sunlight and causing temperatures to plunge. Anything that had survived the fire of the impact now had to contend with the ice of a permanent winter. Plants died because of the lack of sunlight getting through the dusty atmosphere, and this global famine rippled to the top of the food chain. It put paid to the surviving dinosaurs, and cleared a path for the rise of the smaller, foraging mammals that eventually evolved into humans.

Again, it is a narrative not dissimilar to the catastrophism of Halley and Whiston, but instead of the Earth being shattered and reformed by God, the dominant animal species are wiped out, making way for a new dynasty. According to current estimates, asteroids outnumber the planets in our solar system by a billion to one, and although the vast majority are peacefully circling the Sun in safe orbits, there are some in more eccentric orbits that could draw close and cause a threat. Modern survey telescopes and computerised search techniques show that there are no 'dinosaur killer'-sized asteroids coming our way, but Tunguska-sized objects, which could devastate a city, are not yet completely sampled. And we will probably never identify all the Chelyabinsk-sized objects. We must just deal with those as they appear.

How quickly the starry sky above had gone from being the venue for heaven to a bringer of hell on Earth.

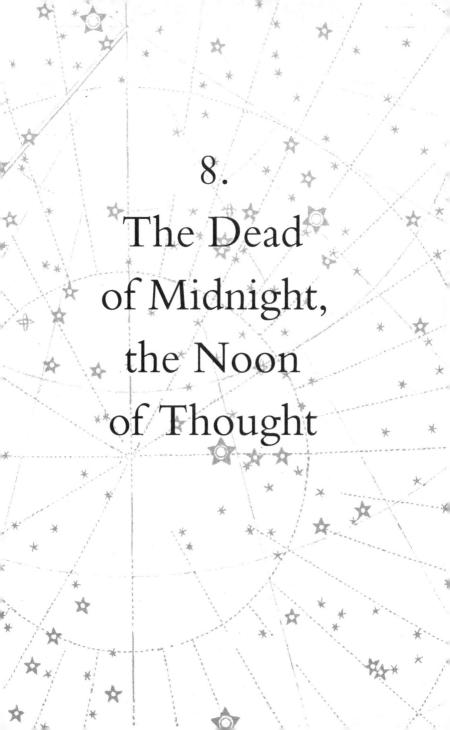

8.

The Dead
of Midnight,
the Noon
of Thought

Before the scientific revolution, the connection an individual felt to the night sky was likely to be immediate and strong. With no light pollution to mar the view, the stars pressed down from heaven and were thought to mirror the comings and goings on Earth. More than that, it was thought that everything on Earth could be understood by studying the state of the night sky. The work of the great astronomers during the sixteenth and seventeenth centuries changed everything. It fatally undermined these old ways of making sense of the cosmos and forced us to begin our relationship with the night sky from scratch; our connection severed except for those few élite individuals, the astronomers, who pursued this new understanding of the universe.

This is the great irony of the scientific revolution. With its emphasis on prediction and proof, the rise of science plunged Europe and its people into a state of profound uncertainty. The old 'truths' about astrology and our links to the night sky were swept away, with nothing to replace them except the promise that science would eventually provide the answers.

In the early twentieth century, while looking back at this moment and how it led to the rise of the modern world, German sociologist Max Weber coined the term 'disenchantment'. He saw the Enlightenment and the scientific revolution as the moments when we stopped enchanting

Nature with talk of gods and spirits, and focused on rationality and science to solve our mysteries. In doing this, he believed, we robbed ourselves of something magical that sparked our imaginations and connected us in an emotional way to the night sky and the other marvels of nature.[70] The loss of this connection, he thought, affected us on a deep psychological level.

One thing that changed immediately was our perception of the night sky. It could no longer be viewed as a firmament, a more or less two-dimensional pattern of twinkling stars and planets. Instead, it was shown to be a three-dimensional realm of vast size that most likely contained many other worlds.

In 1704, Isaac Newton produced a second great book. Called *Opticks*, it was mostly a discussion of the work he had done on light and colours way back in the 1670s. He had only waited this long to publish it so that his nemesis Robert Hooke, who disliked the conclusions, was safely dead and buried. He wrote the book in English instead of Latin and included an extraordinary end section titled 'Queries'.

In his early sixties by this time, Newton had constructed more scientific ideas than he could possibly investigate in the time remaining to him. So he framed these hypotheses as rhetorical questions and listed them in the final section of his book, effectively setting the agenda for the physical sciences that we continue to follow to this day.

The eleventh query related to the night sky and fully addressed the idea that there was no fundamental difference between the Earth and the cosmos. In part, the query said, 'And are not the Sun and fix'd Stars great Earths vehemently hot?' With this somewhat archaic phrase, Newton was proposing that the Sun and the stars are the same type of celestial object. By implication, then, there could be planets around the stars. To take the statement to its logical conclusion, some of those planets could be inhabited like the Earth. This idea was known as the plurality of worlds, and was not universally accepted.

Kepler, for example, had been repulsed by the idea. Like many others, he believed that God lived in the seventh heaven. If so, an infinite universe pushed that realm to an infinite distance away, which Kepler found a terrifying thought.

In contrast, the literary figures and artists of the seventeenth and eighteenth centuries found a vast cosmos full of other inhabited planets much easier to comprehend. Perhaps this was because there was already a long tradition of literary interest in contemplating our emotional responses to the night sky. The strange blend of wonder, with its overtones of pleasure and its undercurrent of fear, proved to be a touchstone for everyone feeling lost in the wake of science's new way of thinking about the cosmos. Despite all that had changed in our understanding, the one thing that remained constant was the feeling of awe that we experience whenever we look up.

So as the astronomers toiled at the reconstruction of our intellectual understanding of the night sky, and how to square that with religious points of view, the artists and philosophers turned their attention to the enduring emotional effects that those twinkling stars evoke.

* * *

Ruminations on the night sky had been a feature of the artistic world since at least the first century BCE, when Roman poet and philosopher Lucretius wrote *De rerum natura* (On the Nature of Things).[71] It was a didactic poem that set out the views of fourth-century BCE philosopher Epicurus, who contended that the cosmos was the product of natural laws rather than the direct intervention of a god or gods.

Epicurus believed in atoms, which he imagined to be the tiny building blocks of all matter. For evidence, he pointed out the way stone steps are worn down through repeated use. The imperceptible nature of the day-to-day erosion proved to him that minuscule bits of matter were being rubbed away with each footfall. He also believed in the plurality of worlds, because he thought the ceaseless jostling of atoms in a vast or infinite cosmos would naturally build planets around the Sun and other stars. These two ideas became joined; to believe in atoms meant to believe in the plurality of worlds, and therefore in a vast cosmos. In contrast, when Christianity arrived, its theologians went the other way and built a narrative based only on what could be seen in the night sky.

As we have discussed, the Church adopted the belief that the stars were fundamentally different from the Sun and represented a tangible boundary (the firmament) beyond which was God's heaven. The Earth was therefore singular, with God's grace being bestowed entirely upon us humans. So to suggest that there were myriad worlds scattered throughout space was effectively a heretical statement.

An example of how dangerous it was to hold this idea can be gleaned from the fate of Giordano Bruno. He was an

Italian Dominican friar, born in 1548. In his twenties, he developed a number of ideas that put him at odds with the core beliefs of the Roman Catholic Church. One of these was that there were other inhabited planets. After years of investigation and a trial that lasted a further seven, Bruno was burnt to death by the Roman Inquisition. Undoubtedly, the severity of his punishment had more to do with his repeated questioning of Christ's divinity and Mary's virginity, but his belief in the plurality of worlds was still important enough to have been listed as one of the charges against him.

In his own work, Newton tried to unite both points of view. He clearly favoured the idea of a vast cosmos, unified by the force of gravity acting the same everywhere, but that did not mean he went along with the Epicurean idea of deposing God in favour of physics. Edmond Halley's foreword to *Principia* makes this clear. Borrowing from Lucretius, he wrote it in the form of a didactic poem.

The very first verse makes no bones about telling the reader that God is responsible for the law of gravity:

> *Lo, for your gaze, the pattern of the skies!*
> *What balance of the mass, what reckonings*
> *Divine! Here ponder too the Laws which God,*
> *Framing the Universe, set not aside*
> *But made the fixed foundations of his work.*

Later in the poem, Halley introduced an idea that became extremely well used by subsequent poets. It was the notion that scientific thinking (referred to here as genius) could allow us to metaphorically journey into the heavens.

Those on whom delusion cast its gloomy pall of doubt,
Upborne now on the wings that genius lends,
May penetrate the mansions of the gods
And scale the heights of heaven. O mortal men,
Arise! And, casting off your earthly cares,
Learn ye the potency of heaven-born mind,
Its thought and life far from the herd withdrawn![72]

The final two lines are notable because they allude to the idea of human beings created in God's image, a feature of Judaism, Christianity and certain forms of Islam. Halley suggests that the similarity is not in the physical form, but in our intelligence and the trained mind's ability to comprehend the surrounding universe. This struck a chord with the poets of the period, who adopted both Newton's concept of the vast universe and Halley's idea of mentally voyaging into the night sky to unlock its secrets. James Thomson, whose output includes the lyrics for *Rule, Britannia!*, wrote a memorial poem to Newton shortly after the scientist's death. It epitomises the idea of gravity as the ruling force of the universe and Newton's ability to ascend through his scientific reasoning.

... [Newton] took his ardent Flight
Throu the blue infinite; and every Star
... at his approach
Blazed into Suns, the living Centre each
Of an harmonious System: all combined,
And rul'd unerring by that single Power,
Which draws the Stone projected to the ground.

This marked another significant change in the way we related to the night sky, because it meant in effect that the living could now 'journey' into the heavens and return with knowledge of how they work. Previously, it was thought that the starry realm had been unknowable to us except at our deaths – as illustrated by the star tables on Egyptian coffin lids, which guided departed souls through the cosmos. Now, however, we can get there whenever we want by thinking about it in the right way. It was another example of how the cosmos was becoming a tangible – or at least knowable – realm. The notion was at once exhilarating and daunting, and this combination of pleasure and fear made it fascinating to philosophers who had long been interested in what they called the sublime.

★　★　★

Discussion of the sublime was a common feature of aesthetics, the branch of philosophy concerned with our appreciation of art and nature, and why we find some things beautiful. At its heart, aesthetics is part of a much larger discussion about the way we interact with nature, and the way our senses provide experiences that drive our emotions. In this debate, the term 'sublime' is reserved for experiences that we find beautiful yet too big to comprehend, or which remind us of how small we really are.

In general, a beautiful object is something that gives pleasure to the viewer. It is usually a small, non-threatening thing, whereas a sublime object is a beautiful thing or vista that is so overwhelming it also generates a sort of fear in us. The night sky was the perfect example of the sublime because of the cocktail of emotions we feel when looking at it. In 1712, the English essayist Joseph Addison described the Newtonian universe as a sublime entity, emphasising the barely graspable immensity of it all.

> *When we survey the whole Earth at once, and the several Planets that lie within its Neighbourhood, we are filled with a pleasing Astonishment, to see so many Worlds hanging one above another, and sliding round their Axles in such an amazing Pomp and Solemnity. If, after this, we contemplate those wide Fields of Ether, that reach in height as far as from Saturn to the fixt Stars, and run abroad almost to an Infinitude, our Imagination finds its Capacity filled with so immense a Prospect, and puts itself upon the stretch to comprehend it. But if we rise yet higher, and consider the fixt Stars as so many vast Oceans of Flame, that are each of them attended with a different Sett of Planets, and still discover new Firmaments and new lights, that are sunk farther in those unfathomable Depths of Ether, so as not to be seen by the strongest of our Telescopes, we are lost in such a labyrinth of Suns and Worlds, and are confounded with the Immensity and Magnificence of Nature.*[73]

In effect, Addison is saying that while the human mind can wrap itself around the idea of the Sun and its collection of planets, trying to imagine the immensity beyond is too much. The final sentence of the passage presents the idea of an uncountable number of planets like Earth that no telescope can see, and it is this that drives Addison into feelings of the sublime. It defines the boundary between sensory experience (we can see the Sun's planets) and the leap of imagination necessary to contemplate the other worlds we feel are there but cannot actually observe. To perform this leap successfully, the imagination is guided by scientific thought, and this marks the night sky out as something unique, something that is now termed the cosmic sublime.

In 1757, the Irish philosopher Edmund Burke wrote *A Philosophical Enquiry into the Origin of Our Ideas of the Sublime and Beautiful.* The book became a standard textbook on the subject, but surprisingly the author only mentioned the night sky once, simply stating: 'The starry heavens, though it occurs so very frequently to our view, never fails to excite an idea of grandeur.' It reads as if Burke is saying the idea of the cosmic sublime is so obvious it needs no further discussion.

The poets of the period fully explored the effect of being beneath the night. In particular, nocturnes were a popular form of poetry that suited the discussion perfectly. They typically celebrated the common human experience of finding it easier for the mind to let go of the tangible world at night-time. When darkness falls, so our thoughts strive towards more contemplative subjects.

Anna Barbauld's 1773 poem *A Summer Evening's Meditation* begins with a quote from Edward Young's 1742 poem *Night Thoughts.* The simple sentence perfectly captures Newton's eleventh query by stating, 'One sun by day, by night ten thousand shine'. Referring to the sun as a sultry tyrant, Barbauld then tells how Venus shines in the twilight evening sky, impatient for the night to arrive. When darkness fully descends, she describes the emotional change this brings in us: 'This dead of midnight is the noon of thought, and wisdom mounts her zenith with the stars.'

In the *Critique of Practical Reason* (1788), the German philosopher Immanuel Kant expanded on a similar idea, making an explicit link between our humanity and our contemplation of the night sky. 'Two things fill the mind with ever new and increasing admiration and awe, the oftener and the more steadily we reflect on them: the starry heavens above and the moral law within.' He goes on to assert that the two things are not separate concepts: 'I see them before me and connect them directly with the consciousness of my existence.'

This connection is the key for Kant because the starry heavens and the moral law within represent oppositely directed human tendencies that must be reconciled. On the one hand, our view of the night sky locates us in a vast cosmos, making us feel small and inconsequential. Kant goes as far as to say that this realisation 'annihilates' our importance as individuals. On the other hand, our recognition of right and wrong and the choice this gives us over our actions elevates us above all the other animals. It also puts us above the dumb matter of the universe, which can do nothing but slavishly follow the laws of physics. In effect, what Kant is describing is the sublime, that strange mix of fear (at annihilation) and pleasure (at comprehension).

There is a particularly beautiful symmetry to the way Kant compares the infinite expanse of space to the infinite depth of thought within each of us. By doing this he is saying that the very measure of being human is our ability to perceive and understand the night sky, coupled with our ability to tell right from wrong. How noble this makes us sound, both as a species and as individuals. He points out that each of us is a finite being, conscious for just a small amount of time, and yet we aspire to understand the infinite cosmos in which we find ourselves alive.

In 1790, Kant wrote the *Critique of Judgement*. In this work, he extended Burke's earlier treatment of beauty and the sublime, and dealt with our contemplation of the infinite universe head on. He expanded upon the idea that while our senses have limitations, our intellect does not. In his view, an object can be beautiful if we can experience it in its totality. For example, a flower can be completely appreciated because we can look at it from every angle, we can touch it, smell it, even taste it if we like. By contrast, the vast universe around us overwhelms our senses, making it impossible for us to experience in totality, and so we experience the sublime. Kant saw that mathematics offered a way to bridge the two experiences.[74]

Newton's law of gravity applies throughout the universe and explains a dizzying array of phenomena. For example, the reason stars and planets form, the reason they stay in orbit, and the reason we stick to the surface of the Earth can all be traced to Newton's mathematical equation for gravity, which depends on just four mathematical quantities and can therefore be understood in its entirety. This is why scientists talk about equations and the theories they represent as being beautiful: because they can be appreciated *in toto*. So Kant argued that mathematical reasoning allows us to develop ways of understanding the things that our senses are incapable of perceiving in their entirety. In one fell swoop, he canonised the abstract reasoning that is now integral to our scientific investigation of things too big for us to experience. And he gave a name to the awesome pleasure we experience when we capture an underlying rule of nature in a mathematical expression. He called it the mathematical sublime.

But in the eighteenth century the vast majority of people were not skilled in mathematics. Only the élite, gifted or lucky would have received a formal education. Even now, many shy away from maths at school. So does this mean that the unique thrill of the mathematical sublime is beyond the reach of most people? Not at all. As the century progressed, a solution evolved. It gave birth to a discipline that straddles the boundary of art and science, and remains widespread to this day. You're currently reading an example of it: science popularisation.

Science popularisation seeks to recreate a feeling of the sublime by stimulating public imagination with scientific knowledge. It guides our minds to think about things that are beyond our ability to experience, yet which can be illuminated with scientific investigation or mathematical analysis. Instead of working through all that mathematics, however, the populariser translates the conclusions into words or images and uses them to stimulate the same imagination that is unleashed when we are beneath the night sky.

An early example was *Conversations on the Plurality of Worlds*, by the French author Bernard le Bover de Fontenelle. Published in 1686, a year before Newton's masterpiece *Principia*, it explored the Copernican idea of the Sun being the centre of the solar system, and the question of whether life would be found elsewhere in the universe. The text was a series of conversations that took place between a philosopher and a nobleman as they walked in a garden at night under the stars.

The public science lecture also became increasingly popular through the eighteenth century, because its practitioners learned how to translate science into theatre. They did it by using experimental demonstrations and other pieces of apparatus, and their own skills in oratory. In effect they blended the literary tradition of the sublime with Kant's mathematical one to present the cosmos in a way that no one had experienced before.

A particularly important prop in these presentations was the orrery. These were devices that used clockwork mechanisms to move a model of the solar system in a way that demonstrated the movement of the planets around the Sun. As the various globes circled at their different speeds, so the viewers could experience a god's eye view of our solar system. In this way, what seemed mathematically abstract to them before now felt graspable and real, thanks to being able to see a model of it with their own eyes.

The artworks of Joseph Wright of Derby, such as his 1766 work *A Philosopher giving That Lecture on the Orrery, in which a Lamp is Placed in the Position of the Sun*, capture the rise of this new pastime. And it wasn't just astronomy that was being presented to the public in this way. Wright's most famous picture is his 1768 work *An Experiment on a Bird in the Air Pump*, that shows a bird floundering as a white-haired philosopher pumps the air out of its chamber; his audience looks on with emotions that run from fascinated to uncomfortable to disturbed.

While these particular paintings show the demonstrations taking place in front of wealthy household gatherings, a wider public thirst for such presentations was growing. As demand grew, so did the venues, the audiences and the equipment. In the north of England, one family struck upon such a winning formula that it allowed them to lecture publicly about the night sky for seven decades. The key to their success was that they did not present their demonstrations as science but as aesthetic experiences, primarily designed to use the newfound knowledge of astronomy to instil feelings of the sublime in their audiences.

* * *

Adam Walker was born in the Lake District of England in 1730. He was the son of a wool trader and possessed very little formal education. By the 1760s, however, he was running a school in Manchester. Following the birth of his first son in 1766, he left his employment, bought demonstration equipment from a retiring lecturer and spent the next several years travelling northern England, Scotland and Ireland lecturing on science.

His work brought him to the attention of bona fide scientists such as Joseph Priestley, who had discovered oxygen in the mid-1770s. Priestley believed that the public should be educated in the ways of science and was a high-profile lecturer himself. He was undoubtedly a big influence on Walker, and in the late 1770s gave him additional demonstration equipment. Both men believed that there were moral advantages to the study and dissemination of science.

In 1799, Walker published *System of Familiar Philosophy*, a popular-level book aimed at sharing knowledge and defending science as a route to personal and societal enlightenment. His career was already successful, but what made Walker famous was his decision to lecture about astronomy. In his quest to bring the grandeur of the night sky and the surrounding universe into theatres he invented a machine that he called the Eidouranion, which comes from Greek and means 'image of the heavens'.

No plans or artefacts remain of this machine, so no one knows exactly how it worked, but it clearly went through a number of iterations, firstly to improve the design and then to make bigger versions that Walker and his sons could use in grander venues. From the many reviews that were published in the newspapers of the time, it is clear that the Eidouranion was a large screen that was suspended on the stage. On the front of it were a series of transparent images of the night sky – for example, the signs of the zodiac and the planets – which were illuminated from behind. When the lights were dimmed in the theatre, the coloured images shone out as if suspended in the darkness of space. But the best part was that as Walker talked to his audience about the wonders of the cosmos, unseen mechanisms whirred into life, causing the planets to orbit the Sun and rotate to simulate day and night. Each planet moved at a representative speed, and together they wove a tapestry of movement that held the audience spellbound.

A description of the machine written in 1782 and published in the *Morning Herald and Daily Advertiser* concluded, 'Besides its being the most brilliant and beautiful spectacle, it conveys to the mind the most sublime instructions.' A much later article, published in 1840 in the *Magazine of Science*, agreed with this conclusion, saying that a similar device 'conveyed some at least of the infinitely more stupendous apparatus of the universe'. Adam Walker's Eidouranion was a successful way of allowing his audiences to share in the feelings of the cosmic sublime partly because it embodied the defining characteristic of the universe itself: mystery. It was completely unclear to the audience how the contraption worked, just as it was an utter mystery to most how gravity worked and turned the universe in its myriad clockwork-like ways.

As the eighteenth century was drawing to a close, the western world found itself riven by conflict. Across the Atlantic was the American War of Independence, and across the Channel, the French Revolution. In addition to the lives these were claiming, they were breaking up the traditional order of things with the loss of Britain's colonies in America and of the aristocratic hold on power in France. Many saw a parallel to the way the new investigation of nature was breaking down the traditional view of the world, and so blamed science and its progressive tendencies for helping to foment these upheavals. This association was strengthened by the fact that some scientists were outspoken in their sympathies for the overseas revolutionaries, and one of these so-called Dissenters was Adam Walker's mentor Joseph Priestley.

In the summer of 1791, a series of riots broke out in the English city of Birmingham. The hooligans targeted the homes of Priestley and other Dissenters and, through guilt by association, the homes of scientists who attended an organisation called the Lunar Society. The ordeal was so terrifying for Priestley that he was immediately driven from the city and eventually left the country.

The emotions behind the backlash are easily identifiable as part of Max Weber's 'disenchantment', and came together in the early nineteenth century as the Romantic arts movement, which sought to concentrate on our individuality and emotional response to nature. In London, however, one commentator who had witnessed the Walkers' astronomy show defended the new study of the night sky. Writing in the *Monthly Mirror* he stated that the veneration of astronomy and the celebration of its ability to unlock the 'beneficent' cosmic design was the opposite of the sentiments that were driving the violent upheavals in the world.

A modern analysis of the various reactions and reviews to the Walkers' show was performed by Jan Golinski, professor of history and humanities at the University of New Hampshire, in 2017. He said in conclusion, 'By conveying the majesty of the cosmos in their lectures, the Walkers elicited quasi-religious feelings of wonder and awe in their audiences.'[75]

It was undoubtedly this reverence that allowed the Walkers to talk about things that had previously been denounced as atheistic. In particular, they championed the idea of the plurality of worlds, using the overwhelming idea of other inhabited planets as a way to demonstrate the sublime majesty of the universe. Whereas before the idea of other inhabited worlds had been firmly linked to Epicurus and his idea of a godless universe, the Walkers jettisoned this association and talked instead of a divine cosmos, thereby implying that God had designed the universe to be filled with planets all along. The inclusion of this religious explanation provided a safety net for anyone feeling too adrift in the vastness of space. The nocturnes of the time often used a similar trick. They explored the ideas of the cosmic sublime by contrasting our physical senses with our scientifically driven imaginations, and then pushing the idea into religious themes in the later verses when it all threatened to become too overwhelming.

★　★　★

Throughout the nineteenth century, science popularisation became more and more established as a means of spreading the scientific view of the cosmos, and a cadre of novelists became interested in a paradox that they noticed this caused in us.[76] When we look into the night sky, we subjectively feel that we are looking at a black dome with twinkling stars placed upon it, and the Earth in some sort of central position. But science tells us that objectively we are looking into a mostly empty void. The occasional pin-prick star is, in fact, a burning mass similar to our Sun – a whole new 'centre' to the universe. Reconciling the natural interpretation our brain puts on the experience with the actual facts of the matter leads each of us to develop a 'picture' of the wider universe that we hold in our imaginations. Anna Henchman, an associate professor of English at Boston University, Massachusetts, calls this mental image 'the starry sky within'.[77]

Some writers of the time drew parallels to the difficulty we have in reconciling subjective experiences with objective facts in our everyday lives, and they began mining the subject of astronomy for metaphors that they could use in writing their novels. These writers, such as George Eliot, Thomas Hardy, Charles Dickens and Leo Tolstoy, peppered their narratives with references to the night sky and the work of astronomers. They took inspiration from the way we adjust to our changing understanding of the night sky by constructing stories that often contrast a central character's subjective experience with the viewpoint of an objective narrator who informs the reader of the whole picture.

These novelists also helped to develop the epic multi-viewpoint novels that we take for granted today. Here we have different characters seeing each other's actions from different viewpoints, but only the reader has the objective overview of what all the characters are thinking. In this way, the reader sees the fictional world as a whole, while simultaneously experiencing each character's individual perception of it.

In the early chapters of Thomas Hardy's 1882 novel *Two on a Tower*, his central character Swithin St Cleve woos Viviette Constantine with tales of how astronomy is unlocking the secrets of the night sky. Hardy revealed in the book's preface that 'This slightly-built romance was the outcome of a wish to set the emotional history of two infinitesimal lives against the stupendous background of the stellar universe.'

St Cleve explains that the stars hide their true natures from our view. Only when we look more deeply with telescopes and other instruments do we see that the serene tableau that faces us from the night sky is a façade. The stars are revealed to be variable, stormy, even tempestuous. They are separated from each other by vast tracts of emptiness that defy easy comprehension – much like the way a person's inner thoughts are hidden from view to strangers.

In George Eliot's final novel, *Daniel Deronda* (1876), the author uses the words and phrases associated with stars and the night sky to talk about the outer reaches of human experience.

Novelists were not the only ones trying to reconcile the new knowledge of the night sky with their innate experiences. So too was the artist Vincent Van Gogh. It was Christmas 1888, and the artist found himself embroiled in an argument with fellow artist Paul Gauguin at their shared house in Arles in southern France. While the details of their disagreement remain unclear, the results were terrible. Van Gogh retreated to his bedroom and severed part or all of his left ear with a razor blade. This set in motion a chain of events that culminated in Van Gogh painting what is arguably his masterwork, *The Starry Night*.

Earlier that year, Van Gogh had written to his younger brother with a confession. Clearly struggling to fill the gap left by the loss of his religious faith, Van Gogh reported his 'tremendous need for, shall I say the word – for religion – so I go outside at night to paint the stars'.[78]

Following his injury, Van Gogh voluntarily entered the asylum of St Paul de Mausole. One day he woke early and looked out at the sky before dawn. The landscape was in silhouette, yet the sky was alight with a waning crescent moon, the bright planet Venus and several stars. Van Gogh painted the scene.

In *The Starry Night* a town with a church lies darkly below the stars. The church's steeple just touches the sky, but in the foreground a cypress tree reaches clear into the nightscape.

In 1845, William Parsons, 3rd Earl of Rosse, (1800–67) built the world's largest telescope at Birr Castle, in County Offaly, Ireland. (Interfoto/Alamy)

The most striking part of the composition, however, is the swirl of faint light that the artist put in the very centre of the image. It appears at first to be a surrealistic flourish, but to any nineteenth-century astronomer, Van Gogh's inspiration would have been obvious.

In 1845, William Parsons, 3rd Earl of Rosse, had built the world's largest telescope. Called the Leviathan of Parsonstown, it was sited at his ancestral home, Birr Castle, in County Offaly, Ireland. The mirror measured seventy-two inches in diameter and was housed in a telescope fifty-four feet long, suspended between two forty-foot-high brick walls.

Using Leviathan, Lord Rosse had sketched an extraordinary swirl of faint light in the northern constellation of Canes Venatici, the hunting dogs. Invisible to the naked eye, this delicate offering is now known to be a distant galaxy containing hundreds of billions of stars. As new stars are born in this slowly rotating system, so they naturally form these beautiful spiral arms. At the time, however, its nature was unknown; it was nothing more than a jewel of the night that tantalised us with its mystery.

Van Gogh, therefore, was simply putting it in its rightful place, perhaps as a talisman of his own attraction to the mysterious promise of the night sky.

What is clear is that in depicting the stars in these ways, the novelists and artists of the period were drawing the night sky and its mysteries closer to us, bringing them into our common experience. They were then using our innate response to these things as a means to illuminate our daily lives. All of it, they were saying, can be rationalised in the same way that we make sense of our competing ideas about the night sky. In essence these writers and artists were exploring the cosmic sublime and the paradox of seeing something yet believing it to be different.

I experienced this paradox first hand in the mid-1990s. I was a research student on an observing trip to the Anglo-Australian Telescope, on Siding Spring Mountain in Australia's Warrumbungles range. I had been awarded my bachelor's degree in astronomy and was now researching for a PhD. By any standard, I was well versed in the theoretical understanding of the cosmos, but the first night I stood under the Australian sky, the sublime truly took hold of me. There was no light pollution on the mountain and I was staggered by the number of stars I could see. There were so many that I found it difficult at first to pick out the familiar constellations. The stars appeared so bright and so 'close' that I felt an almost irresistible urge to reach up and pluck one from the sky. It was strangely disconcerting: one part of my brain knew that the stars were suns at extreme distance, yet another part told me I could take hold of one as if I were some sort of god.

In thinking about the experience later, I decided that such flights of fancy occur when our brains hold two conflicting images of something. Usually one image is very large and intellectually defined, the other subjective and personal. It may even be this kind of juxtaposition that drives our creativity: the desire to encapsulate something overwhelming into something we can wrap our heads around.

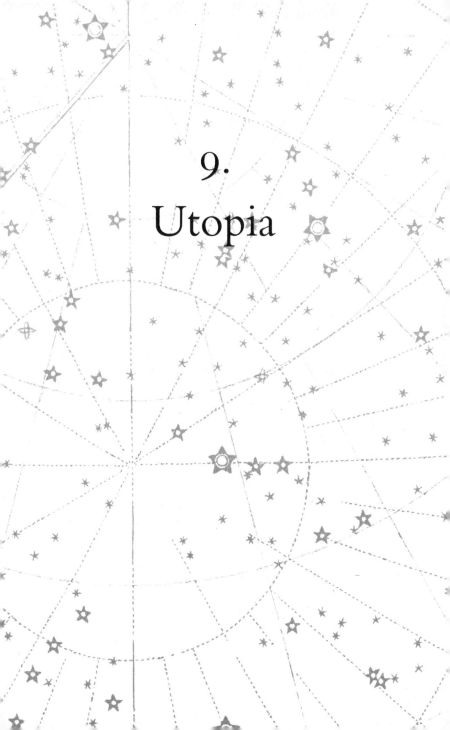

9.
Utopia

As the astronomers revealed more and more about the night sky and the celestial objects it contained, so people increasingly came to think of it as a physical realm, rather than a heavenly one. And that meant people also began thinking about what it would be like to explore it.

From times of antiquity onwards, writers and poets have occasionally exercised flights of fancy by dreaming of travelling up into the night sky. One early example was *True History*, written in the second century CE by Greek satirist Lucian of Samosata to poke fun at travelogues. In the story, he described a trip to the Moon, which was found to be anything but heavenly and just as war-ravaged as the Earth.

In the seventeenth century, when astronomers were making their great telescopic discoveries, more and more of these stories appeared. In 1638, Anglican clergyman John Wilkins published *The Discovery of a World in the Moone*. It was followed two years later by *A Discourse Concerning a New Planet*. Both popularised Galileo's idea of the Moon as a world in its own right. Wilkins also looked forward to the invention of flying machines and then spacecraft, seeing it as a fundamentally human desire to reach previously unreachable places. Also in 1638, another churchman, Bishop Francis Godwin, published *The Man in the Moone* as a fictional exploration of our nearest celestial neighbour.

Bishop Francis Godwin (1562–1633) published *The Man in the Moone* in 1638 to describe what an exploration to our nearest celestial neighbour might be like. The frontispiece depicts the unlikely journey. (British Library)

The frontispiece of the book memorably depicts a flock of swans transporting the central character upwards on his voyage.[79]

Sometimes it was the astronomers themselves who wrote the stories. Johannes Kepler penned a work called *Somnium* in 1608, while he was in the midst of his work on planetary orbits. The title comes from the Latin word for 'dream' and the story follows the adventure of Duracotus, a fourteen-year-old Icelandic boy who is taken to the Moon during a total solar eclipse by a daemon who can only travel in the shadows. Kepler correctly describes overcoming the Earth's pull of gravity using an acceleration, before entering the Moon's gravitational field, which necessitates a deceleration to gently touch down on its surface. In his story, the Moon is called 'the island of Levania' and Earth is called Volva.

Once safely on Levania, Duracotus sees what a solar eclipse looks like from his new perspective. He also experiences the two lunar hemispheres; the near side, which Kepler names Subvolva, and the far side, called Privolva. The Earth is not seen from Privolva, whereas it is always in the sky over Subvolva. Kepler describes how the Earth would go through phases like the Moon, and gives a detailed description of the movements of the Sun, Earth and planets as seen from the Moon. *Somnium* is important because it presents scientific insight in the form of a story. Kepler never sought to publish it himself; it only saw the light of day in 1634, when Kepler's son Ludwig arranged for it to be printed posthumously. Yet for all its astronomical accuracy, Kepler was still forced to rely on magic to transport Duracotus to the Moon in the first place.

The first story that proposed a realistic idea of using rockets for space travel came in the 1893 novella *On the Moon* from Russian polymath Konstantin Tsiolkovsky. He was one of a group of Russian idealists who imagined a bold future for the human race in which outer space was conquered and the Moon and the planets were colonised. Tsiolkovsky had been inspired by a Russian librarian called Nikolai Fyodorov, who in the seclusion of the catalogue room at the Rumyantsev Museum, Moscow, had developed the most radical ideas about human destiny. This extraordinary set of beliefs, which included the religiously inspired goal of returning the dead to life, sparked a utopian philosophical movement known as cosmism, which eventually led to the Russian space programme and the American Moon landings of the 1960s.

Fyodorov was born in the early summer of 1829. His first job was as a teacher of history and geography, but what really inspired Fyodorov was the future. A follower of the Eastern Orthodox Church, he believed that God worked through us to achieve the stated goals of the Bible, which included the development of human immortality through the perfection of science and medicine. Once this had been achieved, he believed we would then be morally obliged to extend that knowledge to the resurrection of the dead.

By doing this, Fyodorov argued, we would be enacting a version of the Bible's final judgement. Through science and medicine he believed we could restore the perfect, God-given state that had been the privilege of Adam and Eve in the Garden of Eden.

Fyodorov saw two reasons for death: illness and accidents. Both, he claimed, could be prevented with sufficiently advanced medicine and science. For example, better knowledge of the environment could protect us from natural disasters. In cases where death was unavoidable, he fell back on his belief that we would eventually develop the science of resurrection. But where would we put everyone? If no one died, Earth would soon run out of room.

This is where Fyodorov looked to the night sky. He saw a connection between the scientific understanding of the universe and the biblical passage that promised entry into the kingdom of heaven. The plurality of worlds was now generally accepted, and that meant a vast universe of other planets was waiting to be explored, conquered and colonised. He believed that was humankind's destiny; our entry into the kingdom of heaven would be our ability to leave the Earth and colonise space.

He pondered these topics for almost twenty years, and although he published nothing in his lifetime (his one book, *Philosophy of the Common Task*, appeared posthumously in 1903) he would conduct long conversations with anyone who came to his library. This earned him a number of influential fans from the artistic and scientific worlds and he developed a reputation as 'the Socrates of Moscow'.[80] Novelist Leo Tolstoy was a frequent visitor, which may explain why so much astronomy and cosmic imagery found its way into his novels.

Another admirer, as we have seen, was Konstantin Tsiolkovsky, born in 1857 in Kaluga, about 110 kilometres south-west of Moscow. Rendered deaf by a bout of scarlet fever when he was ten, he was excluded from school because of his hearing problems and became a recluse. Shunned by his neighbours and townsfolk for appearing odd, he filled his days by reading, gravitating towards mathematics and physics. At sixteen, he went to live in Moscow so that he could read in the great libraries and attend science lectures, where he would use a hearing trumpet. It was during the three years he spent in Moscow that he met Fyodorov and came to share the older man's vision of the future.

Unlike his mentor, Tsiolkovsky had the ability to turn the space travel ideas into scientific theories, and he began to search for a practical way to allow humankind to reach the night sky. Hearing of the construction of the Eiffel Tower in 1895, Tsiolkovsky first imagined a space elevator, a giant metal tower stretching all the way into space. When his calculations showed that no known material was strong enough for the task, he switched to the mathematics of using rockets to climb into orbit.

Tsiolkovsky was prolific. He published more than four hundred works, with just under a quarter of them dedicated to rocketry and space travel. His most important contribution was the Tsiolkovsky rocket equation in which he related the mass of a rocket and its fuel to the velocity it could achieve. Although he wasn't the first to derive this equation, he was the first to use it in connection with a discussion of whether a rocket could achieve the speeds necessary to travel into space. He published the discussion in 1903, the same year Fyodorov's work came out, and then wrote science fiction stories that could illustrate his ideas in more palatable ways for a general audience.

On the Moon tells the story of two men who wake up on the lunar surface. It describes the low gravity on the Moon and the lowering of water's boiling point because of reduced atmospheric pressure.[81] In later works, such as *Dreams of the Earth and Sky* (1895), he depicted how colonies could mine precious metals and other mineral resources on the Moon and on asteroids. He also wrote of how we could construct giant greenhouses in space to grow food. *Beyond Planet Earth* (1920) was the story of an international group of scientists who build spacecraft and habitats in Earth's orbit so that they can then reconnoitre the solar system with a view to establishing colonies.

In all the thousands of pages that Tsiolkovsky wrote, perhaps the best summation of his lifework is contained in his quote: 'Earth is the cradle of humanity, but one cannot live in a cradle forever.' Yet, for all of his erudition and technical precision, Tsiolkovsky's works failed to generate much interest. No one, it seemed, was interested in actually travelling into space. But that was soon to change.

★   ★   ★

At the beginning of the twentieth century, Russia was lagging behind the West. It had been slow to embrace industrialisation, and now it was facing new challenges. Its cities were overcrowded, with poor living conditions for many of the workers that the new industries relied upon. Food was often in short supply, and many blamed the ruling aristocracy. In a sequence of bloody coups, the Tsar was overthrown and Lenin came to power. He seized control of the country in 1917 following the October Revolution and five years later tightened his grip when his Red Army emerged victorious from the resulting civil war. This was the birth of the Soviet Union, which promised that together the people would work towards a better future for all. And in that broad doctrine, cosmism found its home.

Tsiolkovsky's works suddenly became widely available. The Soviet Union's academics got to read his technical works, and the general public his science fiction stories and other popular writings. In 1932, he looked even further into the future with his *Cosmic Philosophy*. Not so much a novel as a discussion, the book described a time when humans had ventured beyond the Sun's planets and ranged freely among the stars.

In this new political era, cosmism struck a deeply resonant chord. The measure of its influence can be gauged by 1935's May Day parade. During these annual celebrations, Soviet leaders and people gathered in Moscow's Red Square to witness a military parade and celebrate the Soviet state. At the 1935 gathering, Tsiolkovsky himself was invited to address the crowd. At seventy-seven and in failing health he spoke confidently about the future:

> Now comrades, I am finally convinced that a dream of mine – space travel – for which I have given the theoretical foundations, will be realised. I believe that many of you will be witnesses of the first journey beyond the atmosphere. In the Soviet Union we have many young pilots … [and] I place my most daring hopes in them. They will help to actualise my discoveries and will prepare the gifted builders of the first space vehicle. Heroes and men of courage will inaugurate the first airways: Earth to Moon orbit, Earth to

Mars orbit, and still farther; Moscow to the Moon, Kaluga to Mars![82]

A few months later he died at his home in Kaluga. By then the Soviet effort to develop rockets was under way, led by Sergei Korolev. Born in 1907, he had become fascinated by aircraft as a child and by the age of twenty-four had helped to found the state-sponsored Group for the Study of Reactive Motion (GIRD) in Moscow. Originally it was thought that rockets could power aircraft, but Korolev was swift to recognise their applicability to space travel. Before he could make much progress, however, another political upheaval engulfed the Soviet Union, and threatened not only to destroy Korolev's efforts but to take his life as well.

In 1936, the Soviet leader Joseph Stalin was convinced that some in his government were plotting against him and he began a campaign to eliminate his political rivals. What started as a purge of the Communist Party rapidly spread to government officials, intellectuals, artists, landlords, academics and scientists as Stalin's paranoia spiralled out of control. In 1938, the Soviet secret police arrived at GIRD. They arrested Korolev and others. All were accused of various trumped-up charges. Confessions were extracted under torture, along with accusations against each other, and all were convicted of being traitors.

The institute's leaders, Ivan Kleymenov and Georgy Langemak, were executed. Korolev too was sentenced to death and sent to prison to await his fate, but the purge was starting to run out of momentum and eventually his sentence was commuted to eight years in a state-run labour camp for intellectuals, the so-called Experimental Design Bureau. While he was there, the Second World War broke out and he was put to work designing aircraft, including a rocket-propelled sled to get aircraft into the air faster.

When the war ended, Korolev was discharged from the labour camp and drafted as a colonel in the Red Army. He was sent to Germany to pick through the remains of the Nazi base at Peenemünde, near Germany's border with Poland, where a whole new form of warfare had been developed: rocket-propelled missiles. The architect of that rocket programme was a man called Wernher von Braun. As the German war machine crumbled, von Braun and his men had been ordered to retreat from the base, but rather than return to the German heartlands or fall into Russian hands, they had managed to surrender to the advancing Americans. The centrepiece of von Braun's rocket programme was the V2, which was the world's first long-range guided ballistic missile. Standing fourteen metres tall, with an operational range of two hundred miles, the V2 was used to attack Allied cities including London, Antwerp and Liège, Belgium, killing some nine thousand people in the process.

The Soviet development of such missiles was a priority for Stalin. Korolev was named chief designer and tasked with making that happen. By 1947, he and his team had worked out how to build a V2 for themselves, and they then set about improving the design. By 1953, they were confident they could build a missile capable of carrying a nuclear warhead all the way to the United States. They named their design the R-7 and, as they began work on turning it into a reality, Korolev simultaneously imagined putting it to a less destructive use: sending satellites into Earth's orbit. His co-worker, Mikhail Tikhonravov, authored a report on how the R-7 could be used for such a purpose, and throughout 1954, Korolev sought funding from the Soviet authorities to build and launch an artificial satellite packed with scientific equipment. But the idea failed to spark any real enthusiasm or money. What no one in the Soviet Union knew at the time was that, across the Atlantic, others were working towards similar goals.

★   ★   ★

After their surrender, Wernher von Braun and six colleagues were secretly transferred to the United States of America on 20 September 1945. Once there, they were expected to train Americans in rocketry as well as continue their own rocket development with American funds. They helped prepare and launch a number of V2s that had been shipped from Germany, and performed studies of new designs that could be used for both military and scientific research. The outbreak of the Korean War in 1950 loosened the American purse-strings, and von Braun led the development of the Redstone rocket, the first large American ballistic missile.

During this time, America was well aware of the military potential of space. They imagined developing spy satellites capable of photographing enemy territories from orbit. But such an aggressive move came with a risk.[83] Any military mission that overflew the Soviet Union could provoke an accusation that the US had violated Soviet national sovereignty. There was no precedent for the ownership of the space above a nation except for the international law regarding terrestrial waters and airspace. Under those rules, a nation could confiscate any vessel that entered its territory without the proper authorisation. Had America simply launched a spy satellite that flew over their territory, the Soviets could have appealed to international law claiming their rights had been violated. Clearly, this was something that President Dwight Eisenhower wanted to avoid at any cost.

At a major summit between the Cold War superpowers in 1955 at Geneva, Switzerland, he suggested to his Soviet counterpart the concept of 'freedom of space'. He explained that the development of intercontinental ballistic missiles carrying nuclear warheads by both sides brought with it the fear of surprise attacks. His solution was to suggest that they removed the territorial boundaries in space so that both sides could deploy spy satellites designed to reassure themselves that no aggressive moves were being planned. The Soviets rejected the proposal immediately, thinking it an American ploy to identify targets for missile strikes. So Eisenhower needed a different approach. As luck would have it, the perfect opportunity was just around the corner.

An international committee of scientists were planning a year-long collaboration to study our planet as a whole. Called the International Geophysical Year (IGY), and beginning in 1957, the endeavour included studying phenomena such as the northern and southern lights that occurred at the boundary of the Earth's atmosphere with space. Crucially, the Soviets were planning to join in with the IGY, and given that the best place from which to observe the whole Earth was clearly outer space, Eisenhower saw his opportunity.

On 29 July 1955, just eleven days after the Geneva summit, Eisenhower's press secretary James Hagerty announced that the US would launch the world's first artificial satellite as part of the International Geophysical Year. The spacecraft would carry a payload of scientific instruments to begin humankind's study of Earth from a completely new vantage point. The President gambled that no one could object to a scientific craft circling the Earth on behalf of an international collaboration that included the Soviets. Once the satellite was up and running, it would naturally establish space as an international place, beyond the confines of national territories.

In the Soviet Union, the American declaration made the authorities look again at Korolev's scientific satellite proposal. On 30 August, the chief designer presented an updated report to Moscow's Military-Industrial Committee and the Soviet Academy of Sciences. He promised them a one-and-a-half-ton scientific satellite that would be launched between April and June 1957, just before the International Geophysical Year was scheduled to start, thus upstaging the Americans. Both bodies backed the proposal and the satellite was codenamed Object-D. No public statement was made.

Korolev went to work, driving himself and his teams mercilessly to bring the R-7 rocket to fruition, and adapt it for space. In the US, Eisenhower rejected the use of von Braun's Redstone rocket to launch their satellite and chose a rival being developed by the US Navy, codenamed Project Vanguard. It would prove to be a mistake.

As 1957 arrived, both Soviet and American teams were beset with technical issues. In the US, Project Vanguard was running wildly over budget having ballooned from a $20 million programme to $110 million. Eisenhower complained that the scientists had let their imaginations run away with them, designing larger and more elaborate satellites than he had originally approved. He was adamant that being first to launch was the primary objective, not the quality of the science that would be returned.

In the Soviet Union, they were coming to the same conclusion. They too had run into trouble because the R-7 wasn't providing the thrust necessary to lift their hefty satellite into orbit. Pragmatic as ever, Korolev sought to change the payload. Instead of a fully tooled-up scientific mission, he proposed a small radio transmitter that could be tracked by amateurs across the planet. It would be housed in a metallic sphere, just fifty-eight centimetres across, and would transmit from four trailing antennas. His new plan was approved, the satellite was constructed and a possible launch was pencilled in for 17 September 1957, which appropriately would be Konstantin Tsiolkovsky's birthday. In the end, the date slipped to 4 October.

The launch site was in Kazakhstan, a site that has since grown into the Soviets' premiere launch complex, the Baikonur Cosmodrome. After a tense day of final checks, the rocket – which had been named Sputnik – took to the skies at 10.28 p.m. Moscow time. A crowd of engineers who had worked on it were outside watching as the rocket lifted itself into the dark night sky. Higher and higher it flew until it disappeared from sight. Then the crowd raced for the radio station to wait for the 'bleep bleep' that would signal that the little spacecraft inside (Sputnik-1) had separated from the rocket. They heard it but only for a few minutes. The signal cut off as the speeding satellite dipped below the horizon. This was expected but it still made them nervous. Had the satellite truly made it into space? There were doubts.

They had programmed the rocket to put Sputnik-1 into an elongated orbit of 223 by 150 kilometres. This would mean that the satellite took just over one hundred minutes to orbit the Earth, but as they analysed the data radioed back from the rocket during its ascent, they saw that a malfunction meant the orbit was not going to be as large as they had planned. Anxious now, Korolev decided not to call the waiting Soviet premier Nikita Khrushchev until he was sure of the fate of his satellite. That meant waiting for the signal to restart as Sputnik-1 climbed above the opposite horizon, on the way to completing its first orbit. The engineers settled in for an agonising wait.

About an hour and a half later, the electronic heartbeat again sounded through the radio receiver, and everyone breathed a sigh of relief. Sputnik-1 was in orbit. The place erupted in celebration. From a quieter corner, Korolev made his call to Khrushchev. And the world's relationship with the night sky changed yet again.

* * *

The Telegraph Agency of the Soviet Union (TASS) sent word of Sputnik's success around the globe, and it soon dominated headlines. London's *Daily Express* coined the term 'space age' for the general public in their headline: 'The Space Age is Here'.[84] Opinion columns and editorials eulogised the dawning of this new age and everyone was urged to join in by tuning their radio sets to listen to the signal from space, or to go outside and look for the bright spot of the satellite as it crossed the night sky.

In truth, the satellite itself was so small that it was on the limits of naked-eye detectability, but Korolev knew that the bulk of the rocket would also make it into orbit. At twenty-six metres, this was much larger than Sputnik-1's fifty-eight centimetres, so he had engineers fit reflective panels to the rocket, to bounce sunlight down to Earth, making it almost as bright as the brightest stars in the night sky. Crossing the sky from horizon to horizon in a matter of minutes, it was unmissable, and eerie. It appeared as a single point of light, almost indistinguishable from a star. Yet the unnatural speed at which it moved made it stand out as completely different. For the first time in human history, we were changing the appearance of the night sky, and the reaction this prompted was visceral, even primal.[85]

At that time Homer Hickam was a fourteen-year-old American, growing up in West Virginia. He ended up a NASA engineer who worked on the Apollo Moon landings. In his memoir *Rocket Boys*, which was adapted into the 1999 feature film *October Sky*, he remembers watching the Soviet rocket cross the sky: 'I stared at it with no less attention than if it had been God Himself in a golden chariot riding overhead. It soared with what seemed to me inexorable and dangerous purpose, as if there were no power in the universe that could stop it.'[86]

He was not the only one to feel humbled by the sight. America's leaders were aghast. They had been publicly humiliated by the Soviets in an endeavour that they had subconsciously come to think of as American destiny. Space exploration had been hard-wired into the post-war American consciousness, and a whole genre of literature had grown up around it.

The stories of this era, now referred to as the 'golden age of science fiction', often explored the idea of extraterrestrials, as belief in the plurality of worlds had become universal. They also investigated the possible ramifications that advanced technology could have on our lives. Some of these stories looked into the far future when travel to the stars would be as easy as a voyage across an ocean. Others concentrated on our fledging exploration of the solar system in the near future. In this genre, one author dominated all others: the English writer Arthur C. Clarke. In a similar vein to Tsiolkovsky, he based his science fiction on technically plausible advances that were widely expected to be made during the twentieth century.

Clarke and his contemporaries, such as American novelists Isaac Asimov and Robert Heinlein, entertained their audiences with stories of space exploration that were based on the underlying philosophy of positivism, which only valued things that could be scientifically measured and verified. It defined progress as the application of technology to achieve goals that were otherwise impossible, into which framework the exploration of space fitted neatly.

Von Braun's support of all this aligned him perfectly with American values at the time. So despite some commentators' questions about his role as a Nazi officer in the Second

World War, he became a recognisable – even famous – figure in promoting space exploration. He was also a contributing author to a series of articles entitled *Man Will Conquer Space Soon!* in the magazine *Collier's Weekly*. These articles were illustrated by Chesley Bonestell, who also became a major influence in the American public's perception of space exploration because of his evocative depictions.

Bonestell was born in California in 1888, and developed a love of astronomy in his teens after seeing Saturn through a telescope during a public observing evening at the Lick Observatory, San Jose. He studied architecture at Columbia University, New York City, and went on to paint illustrations of the proposed Golden Gate Bridge in San Francisco to allow potential backers to see what they were putting their money into. He then worked in Hollywood, creating photo-realistic illustrations to serve as backdrops in films such as 1939's *The Hunchback of Notre Dame* and 1941's *Citizen Kane*.

In 1944 his eureka moment came when he began combining his art with his interest in astronomy. Returning to his love for Saturn, he produced a series of paintings that accurately portrayed how the giant planet would appear if one were standing on its different moons. His precise style gave the impression that these paintings were actual photographs and they caused a sensation. Before NASA's robotic spacecraft of the 1970s and 80s, Bonestell's visions shaped the public's idea of what being in space would be like.

His style is notable because he was extending a tradition of American landscape painting that derived from Burke and Kant's ideas of the sublime. Throughout the nineteenth century, artists such as Thomas Cole, Albert Bierstadt and Frederic Edwin Church had painted the most dramatic landscapes that North America had to offer. The settings they depicted were often used to evoke a sense of the frontiers that Americans were advancing into and were painted in such a way as to provoke awe. Bonestell did exactly the same in his paintings of moons and planets. With no real pictures or other references to go on, Bonestell imagined rocky peaks and huge alien vistas that mirrored the terrestrial visions of his nineteenth-century predecessors. When it came to the Moon, he would increasingly emphasise the overwhelming nature of these landscapes by showing tiny human figures and even whole rocket ships nestled into the vista. In creating such images, Bonestell had found the perfect way to symbolise Kant's idea of the cosmic sublime.

During this decade, the way humans associated themselves with the night sky was changing rapidly: it was no longer about sitting under the stars and contemplating them, it was about claiming them, turning them into a human realm. Both America and Russia thought this was their destiny. So the success of Sputnik-1 was a hammer-blow to American pride that destabilised their assumption of global leadership.

On 5 October 1957, the *New York Herald Tribune* called it a 'grave defeat for America'. The battle between communism and capitalism was being played out in the night sky. Feeding on this ideological war, the TASS statement claimed that 'the present generation will witness how the freed and conscious labour of the people of the new Socialist society turns even the most daring of man's dreams into reality'.

America's plight was deepened by an anonymous source from Huntsville, Alabama, who told the Associated Press that he was 'angry and distressed' because the US could have launched first if only its leaders had authorised the use of the proven Redstone rocket, not the trailing Vanguard. Paul Dickson, author of *Sputnik: The Shock of the Century*, believes that this anonymous source was von Braun himself, striking back at Washington for not trusting him.

In response to the crisis, the Americans accelerated work on Project Vanguard. Instead of launching a scientific satellite for IGY, they quickly put together a test satellite called TV3. It was almost identical to Sputnik-1, and launch was set for 6 December, just two months later. A month before that day arrived, however, the Soviets upstaged them again. Sputnik-2 reached orbit on 3 November 1957, carrying the dog Laika, the first living thing to leave the planet.

On the day of Vanguard's launch, film cameras were set up to capture the historic moment of America's entry into the space race. The country watched as the countdown reached zero, and the rocket roared into life. People held their breath as Vanguard began its long climb towards orbit, but just 1.2 metres off the ground the rocket lost thrust. It fell to Earth in what looked like slow motion, triggering an enormous fireball. News coverage was savage, pouring scorn on the attempt by calling the US satellite Kaputnik and other variations. At the United Nations, the Soviet delegation baited their America counterparts by asking if they wished to receive aid from the 'underdeveloped countries budget'. Smarting from the national humiliation, the *New York Herald Tribune* wrote, 'The people in Washington should damn well keep quiet until they have a grapefruit or at least something orbiting around up there.'

That 'something' finally reached space on 31 January 1958, thanks to the Redstone rocket designed by von Braun. It lofted the promised scientific satellite, Explorer 1, into orbit. Having yielded to the German scientist's expertise, the United States was back in the space race, and Explorer 1 made a bona fide important discovery: that Earth was ringed by radiation belts, which could pose a danger to astronauts and electronics if proper precautions were not taken.

Determined not to be upstaged again, the US established the National Aeronautics and Space Administration (NASA) and tasked it with conducting an urgent and rapid expansion of US space exploration. Its goal was to establish the nation as leaders in space technology and exploration. But not everyone agreed this was a necessity. A substantial number of the general public felt that space exploration was a sterile endeavour that placed technological achievement over basic human needs on Earth. They rejected authority and championed individual freedoms. And they certainly didn't need to travel into space to appreciate the night sky. As their ideals gained widespread popularity, they contributed to first the counterculture of the 1960s and then the New Age spirituality of the 1970s.

Yet these movements did not just spring up out of nowhere to oppose NASA. They were the development of a nineteenth-century groundswell that sought to restore what many perceived to have been lost during the Age of Enlightenment. They thought science and technology was a completely wrong turn and that the night sky was something that should be contemplated from a distance. So they set about re-enchanting the night sky with a revival of interest in astrology.

10.

To
Touch
the
Night
Sky

In his poem, *When I Heard the Learn'd Astronomer*, the nineteenth-century American poet Walt Whitman captured the feelings of disenchantment that many in the western world were feeling at the time. He wrote:

When I heard the learn'd astronomer,
When the proofs, the figures, were ranged in columns
before me,
When I was shown the charts and diagrams, to add, divide,
and measure them,
When I sitting heard the astronomer where he lectured with
much applause in the lecture-room,
How soon unaccountable I became tired and sick,
Till rising and gliding out I wander'd off by myself,
In the mystical moist night-air, and from time to time,
Look'd up in perfect silence at the stars.

In Whitman's view, the scientific emphasis on measurement and precision had severed our emotional connection with the stars. In the poem, he suggested that the way to reconnect was simply to forget the maths and stand beneath the night with an open mind. He did not negate the work of the astronomers, he simply acknowledged that there were two ways of looking at things. However, to some in the nineteenth century, this was too lenient a viewpoint.

They wanted to discard the conclusions of the scientific revolution and re-enchant the night sky with fantastical ideas that acted beyond the reach of physics. One of those people was Helena Blavatsky.

Born in 1831, in Yekaterinoslav, which was part of the Russian Empire but is now in Ukraine, Blavatsky came from a privileged background; her grandmother was a Russian noblewoman and her father was a member of the German aristocracy. When her mother died of tuberculosis in 1842, aged just twenty-eight, the nearly eleven-year-old Blavatsky was sent to Saratov, a major Russian city on the River Volga, to be brought up by her grandparents. It was there in her teenage years that Blavatsky became interested in so-called esoteric knowledge.

Esotericism encompassed a range of beliefs that hinged on there being something magical driving nature. Whatever this might be, it lies in a spiritual realm beyond our ability to measure. It is metaphysical, operating beyond the normal confines of cause and effect. To those who believe in esoteric knowledge, the physical universe is often thought to get in the way because it obscures the true spiritual reality.

There has been a long-running interest in esoteric knowledge throughout history. It emerged in the classical world alongside Christianity, and aimed to inhabit a middle ground between the strict rationalism found in philosophy, and the faith demanded by religion. The term itself was coined in derision by the Ancient Greek satirist Lucian of Samosata.

Blavatsky claimed that she learned of the subject from books in the library of her great-grandfather, who had been a freemason in the 1770s. Many freemasons, her grandfather among them, were also Rosicrucians, which was a spiritual order that developed as a reaction to the Age of Enlightenment. It embraced the esoteric disciplines that science had discarded: alchemy, magic and astrology. Rosicrucianism was itself part of a more ancient tradition known as hermeticism, which was centred on the micro-cosm—macrocosm doctrine of 'as above, so below'. Named after the mythological Hermes, the messenger of the gods, the hermeticists believed in a version of astrology that claimed the planets influenced conditions on Earth but did not directly dictate our actions. This meant a deeper con-templation of nature was needed to understand how these influences from the night sky worked.

Blavatsky later said that she had been guided during her time of discovery by a mysterious Indian man who appeared to her in visions. She also said she learned to project her-self through the 'astral plane', a supposedly unseen realm of 'light' that existed between Heaven and Earth, and which constituted true reality. It was said to share space with the 'astral spheres' of the planets, yet to be impossible to see with our eyes. It was the home of angels, demons and spirits.

The next twenty-five years of Blavatsky's life are difficult to untangle because she liberally peppered her accounts with inconsistencies, but the core of the story is that she went on many great journeys to learn of more spiritual ways of life. Her ideas were often derived from eastern religions and philosophies, and she claimed to have encountered many adepts, including magicians, shamen and mediums. She asserted that her own paranormal abilities developed to such an extent that furniture moved of its own volition when she entered rooms. She finally learned to control her powers in 1864, after waking from several months in a coma that had been brought on by falling from a horse. She then stated that she entered Tibet (which would have been an exceptional feat because the country was closed to Europeans in the nineteenth century), where she was taken to a monastery and taught a secret language, so that she could read secret manuscripts and learn the ancient knowledge. Although most historians and biographers believe her stories to be nothing more than flights of fancy, her ideas eventually influenced millions of people across the world.

In the 1870s, the Victorians were in love with the concept of spiritualism. Séances were wildly popular and held up as evidence of a spiritual plane, which was usually thought to be synonymous with the astral plane. When Blavatsky visited America (for real this time) she met Henry Steel Olcott, a reporter who was interested in spiritual phenomena. He fell completely under Blavatsky's spell and began to promote her work. In 1875, they founded the Theosophy Society, meaning divine wisdom. Eschewing the political and aristocratic hierarchies of the past, it espoused equality for all, and a future built on spiritual enlightenment.

In 1875, Blavatsky wrote a book, *Isis Unveiled*, outlining theosophical principles and introducing the term occultism, to mean hidden knowledge. Her work came in for immediate criticism because she had copied large chunks of it from other esoteric publications without crediting her source material. Olcott sprang to her defence, claiming that she did not have access to these other books and implying that any similarity must have been because Blavatsky had spiritually tapped into the fundamental truth that they described.

Putting aside her ethics as an author, *Isis Unveiled* is a grand synthesis of the esoteric tradition and the mysterious 'truths' it attempts to grasp. It is clearly based on Plato's ideas of a hidden, perfect reality. Originally, Plato's reality was a perfect mathematical realm, but to Blavatsky it was a non-corporeal dimension of spirits. And instead of Plato's assertion that rational thought can bring us closer to its understanding, Blavatsky echoed the hermeticists and imagined that her hidden reality could only be accessed in a spiritual way.

Blavatsky said that theosophy resurrected ancient wisdom that had once been widely known yet had been lost for millennia, and that misinterpretation of this knowledge had resulted in the many different religions of the world. She also believed that the western world of the nineteenth century had reached a nadir because of its obsession with the physical, measurable universe. The only way out of this intellectual cul-de-sac, she said, was to concentrate on more spiritual enlightenment, and she championed astrology as a way of accessing such spiritual wisdom from the astral plane.

Astrology had been in the doldrums since the seventeenth century, when it had been rejected by scientists because of the lack of direct evidence that they could measure. Blavatsky defended it by repeating Ptolemy's assessment that astrology was infallible but that its interpretation was difficult, and this led to errors. Similar to the analogy that Thomas Hardy drew between astronomy unlocking the behaviour of the stars and getting to know the mind of another person,

Blavatsky likened astrology to the newly emerging science of psychology, saying that in both disciplines one had to step beyond the 'visible world of matter' to see the real truth.

Blavatsky's lifetime coincided with the rise of psychology as a scientific discipline. In the 1880s, Sigmund Freud began the work that would lead to his invention of psychoanalysis. Freud believed that people behave in ways that are driven by their unconscious mind. By bringing these unconscious thoughts, feelings, desires and memories to the surface through therapy, a person can understand why they behave as they do, and become more objective about the decisions they make.

Blavatsky had taken a parallel approach with astrology, saying that the position of the planets could reveal the reasons behind the often random events of our lives. Back in the second century, Ptolemy thought that the astrological influence was transmitted to us through weather and the forces of nature. Now that the scientists had disproved that, Blavatsky suggested a link that was by definition unmeasurable. In keeping with the idea of 'as above, so below' she suggested that the planets' positions were not in themselves responsible for our actions and the events that befall us. Instead, the planets are responding to the same spiritual forces that surround and influence us. In other words, they are indirect probes of this spiritual energy.

In suggesting this mystical link, Blavatsky was appealing to the feeling of the sublime that we get when contemplating the night sky. She insisted that by reading the planets' positions the skilled astrologer could determine the 'tendencies' of Earth, which were in some ways like the unconscious desires in psychology. Despite its scientific implausibility – or perhaps because of it – *Isis Unveiled* was a commercial success. Theosophical lodges sprang up across the world. By 1885, there were more than 120 lodges in existence. The vast majority of them were located in and around the Indian subcontinent, where a lot of theosophy's ideas were already widely believed by the population.

Blavatsky attracted even more controversy because of her claims of paranormal abilities. These had already been met with scepticism, but when two former employees of the theosophical society in India claimed to have proof that she was a fraud, the scandal hit the English national press. The pair claimed that letters, which magically appeared during her séances, were being fed by accomplices through trap doors. When the accusations were published in *The Times* of London, a member of the recently formed Society for Psychical Research left London for India to investigate and concluded that Blavatsky's séances were nothing more than elaborate parlour tricks.[87]

Weathering the storm, she settled in London and continued to publish books on everything from the existence of the physical universe to the evolution of human beings. All of her work from this period had a strong emphasis on astrology, which may have been what drew the British astrologer Alan Leo to theosophy in 1890.

Born William Frederick Allan in 1860, Leo is a seminal figure in the modern interest in astrology.[88] A cornerstone of his re-invention of astrology was his rejection of the complexity of Ptolemaic astrology, in which the positions of the planets, the Sun and the Moon had all to be assessed. Instead, he vigorously promoted the idea that the Sun held the biggest influence, and that astrology was a tool for character analysis, rather than a means of predicting future events. In his view, the constellation in which the Sun is found at your birth dominates your personality. So, all those born under Sagittarius possess similar personality traits, as do all those born under Pisces, Aries etc. This is the moment when the modern concept of the astrological 'star sign' was born. This is why in 1885 he changed his name to Leo to reflect the constellation that the Sun was crossing at the time of his birth.

In 1898, Leo founded *The Astrologer's Magazine*, which he retitled as *Modern Astrology* a few years later. To attract readers, he offered a free personalised star sign reading with each subscription, so starting the trend for the newspaper horoscope phenomenon that still persists today. Like Blavatsky, he attracted criticism but the attacks on him as a charlatan – both from scientists and other jealous astrologers – only increased his burgeoning fame.

By 1915, Leo had published some thirty books on his re-imagined version of astrology and its links to theosophical thinking. His influence was indeed considerable, both directly and indirectly. When the composer Gustav Holst became interested in astrology in 1913 while holidaying in Majorca with a group of artistic friends, it was Leo's book *What is a Horoscope?* that he picked up and read. So empowering was Leo's prose that Holst even felt confident enough to start casting horoscopes for his friends. And out of this experience came the idea to compose the now famous suite of music, *The Planets*, which captured the supposed astrological influences of the planets on our personalities.

As interest in astrology blossomed again, Blavatsky and Leo saw it as a move towards a more spiritual way of knowing ourselves, one that reconnected us with the night sky and prepared us for the demise of the physical realm and the coming of Blavatsky's prophesied spiritual age. They even felt confident enough to use astronomy to put a date on the beginning of this new age.

In her theosophical writings, Blavatsky had drawn attention to the phenomenon known as precession. We first encountered it in chapter 2. It is a slow wobble of the Earth that makes the direction of north appear to trace out a wide circle in the night sky over the course of 25,772 years. As the direction of north changes in the sky, so does the orientation of the celestial equator – the projection of Earth's equator into space. Its change of orientation is important because it influences when we experience an equinox. The Sun's apparent passage through the sky, caused by the Earth's movement through its orbit, is called the ecliptic. Over the course of a year, the Sun appears to pass in front of each zodiacal constellation in turn until all twelve have been visited and the circuit starts again. The celestial equator is set at a 23° angle to the ecliptic, which means that these two great circles cross at just two places. When the Sun is located at one of these places, called a node, we experience an equinox: equal lengths of day and night.

As the north pole precesses, so the celestial equator moves and the location of the equinoxes creep around the zodiac. The calendar is calculated to take this into account, meaning that while the equinoxes always take place around 21 March and 21 September, the constellation the Sun is passing through on these dates changes gradually over thousands of years. Latter-day astrologers gave special significance to each of these crossings, calling them astrological ages (as opposed to the great ages we met in chapter 6, which had faded from consideration by the twentieth century). But the astrologers disagreed on how best to calculate the length of an astrological age. The simplest way is to say that the full precession cycle takes almost 26,000 years and there are twelve constellations, so a crude estimate is to divide one into the other, which means each astrological age lasts roughly 2,166 years. According to this system, starting in roughly 2166 BCE, the spring equinox took place in the constellation of Aries. Each year since, the exact position moved a little further on until around CE 1, when it crossed into Pisces, and where it remains to this day. It will cross into Aquarius some time around CE 2166.

Individual astrologers often had their own way of defining the constellation boundaries, and therefore reckoning the date of the new age, and according to Leo's calculation, it was not centuries in the future, but mere decades. He claimed that the new age would dawn on 21 March 1928 – making it urgent for individuals to prepare. This coming new age of spiritual enlightenment began to be referred to as the age of Aquarius.

★　★　★

The early twentieth-century rebirth of astrology attracted the attention of psychoanalyst Carl Jung. Perhaps it was Leo's contention that astrology was really a tool for character analysis, or perhaps it was a simple fascination with the revival of popular theories of our supposed link to the night sky that piqued the Swiss psychoanalyst's interest. Whatever it was, he found it a revelation.

In a letter dated 8 May 1911, he wrote to Sigmund Freud, 'There are strange and wondrous things in these lands of darkness. Please, don't worry about my wanderings in these infinitudes. I shall return laden with rich booty for our knowledge of the human psyche.' And indeed he did. His conclusion was that the night sky is the perfect psychological mirror, capable of reflecting our most hidden thoughts.

Jung's ideas are an extension of Freud's notions about how the human mind worked. Freud believed that our behaviours are driven by unfulfilled or repressed desires that fester in our unconscious minds, colouring the way we think. At birth, he said, each of us is a blank slate; the frustrations and resentments build up only later as we live our lives. Jung thought this too simplistic and believed that our unconscious mind is already loaded with ideas and concepts at birth. He called these the archetypes; they were the psychological equivalent of instinctive behaviours.

Jung began developing these ideas because he noticed running themes in his patients' descriptions of their dreams and fantasies. He was also struck by the recurrence of themes and symbols in different religions and myths. Whereas Freud thought our experiences shaped us, Jung believed that they merely modified how the archetypes manifest themselves in our thinking. Look hard enough, he said, and you would always see the evidence of basic archetypes at work.

In his book *Archetypes and the Collective Unconscious*, he listed twelve major archetypes. Some relate to ourselves, such as the Persona, which is the conscious image of ourself we present to the world; the Self, which is the sum of who we really are; and the Shadow, which is all the negative impulses that we suppress. Other archetypes relate to other people, such as the mother, who is a nurturing presence; the sage, who imparts wisdom; and the trickster, whom we cannot trust. Jung believed there were many more archetypes beyond these examples and that in creating the zodiacal constellations, humans were simply projecting archetypes onto the stars, just as they did when they developed myths or religions. In his book *The Structure and Dynamics of the Psyche*, he wrote, 'The starry vault of heaven is in truth the open book of cosmic projection, in which are reflected ... the archetypes.' His conclusion was that in studying astrology, he was learning about the structure of the human mind from the core myths that it contained. He wrote, 'Astrology is a naively projected psychology in which the different attitudes and temperaments of man are represented as gods and identified with planets and zodiacal constellations.'[89]

Although Jung marshalled a wealth of evidence to support his idea, Freud hated it so much that it led to a schism between them. In the decades since, there has also been much criticism of Jung's archetypes, but his basic idea of humans sharing some key concepts and then projecting their individual hopes and fears into the night sky is clearly evident when comparing Blavatsky's spirituality with Fyodorov's advocacy of spaceflight.

While it may seem that Blavatsky's theosophy and Fyodorov's cosmism are implacably opposed ideologies, if viewed from a Jungian perspective they can be seen as almost identical because both are chasing the same thing: paradise, or the coming of a new age. According to Jung, the search for paradise is one of the key archetypal desires, and in these particular manifestations of it, the two competitors both placed paradise in the night sky (just as religion had previously done). The only difference between the new philosophies is that Fyodorov believed only matter existed, whereas Blavatsky believed in a spiritual plane obscured by the material one. And this one key difference gave rise to their wildly different visions for how to make progress towards paradise.

Fyodorov believed that the perfection of humanity required the manipulation of matter: first through the use of technology and then by altering ourselves in what we would now call genetic engineering. In this way we could achieve immortality and spread throughout the infinite universe, colonising other planets as we went. Blavatsky thought humankind's problems were the result of our concentrating too much on the physical. She claimed that it was only by developing the spiritual, 'unseen' side of humans that we could truly understand the universe. Both of these viewpoints, or subtle variations thereof, were adopted by different swathes of the population during the space race. They offer an almost textbook example of Jung's idea that humans project their own hopes and fears into the night sky, and they came to a head in the controversy that surrounded sending the first humans into space in the 1960s.

★ ★ ★

Yuri Gagarin was the communist dream come true. He was born in the isolated village of Klushino, two hundred kilometres from Moscow, on 9 March 1934. His father was a carpenter, his mother was a milkmaid, and on 12 April 1961, aged twenty-seven, he became the first human in space.

Both the US and the Soviet Union had started human spaceflight programmes in 1959. It was inevitable that humans would be sent into space, because after all this time of just looking at the night sky, now the opportunity had arrived to actually experience being in it. Even if not everyone could go, at least we could send a lucky few who could then relay their experiences to us.

NASA selected seven astronauts for its Project Mercury and presented them to a packed press hall in Washington DC on 9 April 1959. The USSR also chose cosmonauts but did so more quietly, spending longer on its decision by first whittling down a shortlist from twenty to six, and then making its final selection on 30 May 1960. Gagarin was in this group, having joined the air cadets in 1951 when he went to study tractors at the Saratov Industrial Technical School. As a cadet, he showed such natural talent for flying that when he was drafted into the Soviet military, he was trained to fly MiG-15 jet fighter planes. He excelled at this too. Meanwhile, Korolev's team was busy designing and testing the Vostok space capsule that would keep him alive on his spaceflight.

In America, NASA was making progress on its Mercury capsule. Learning from their previous mistakes, they placed

The chimpanzee Ham is seen prior to his 31 January 1961 testflight in a Mercury-Redstone rocket. (NASA)

von Braun in charge of the Redstone-Mercury rocket that would launch it. On 31 January 1961, the German launched a test vehicle on a suborbital flight that lasted sixteen minutes and thirty-nine seconds. The trajectory would allow the capsule to coast into space and then naturally fall back to Earth. What made the test significant was that it carried a passenger: the chimpanzee Ham.

Prior to the flight, Ham had been trained to push buttons in response to seeing a flashing light. At the end of his mission, Ham's capsule splashed down in the Atlantic Ocean, where he was retrieved by the US navy and found to be in excellent health except for a bruised nose. He was around four at the time of his flight and lived for another twenty-two years. The data recorded from his mission showed that his reactions to the flashing light had been impaired by only a fraction of a second, and so NASA concluded that it would be possible for a human to control instruments while flying in space. Ham's success paved the way for a human flight. Alan Shepard was chosen as the astronaut to fly and preparations began in earnest. It was an exciting time for all concerned, but before they reached the launch pad, Korolev pressed the button that sent Gagarin into space, and America lost again.

The launch took place on 12 April 1961. Gagarin stayed in radio contact with Earth at all stages of the journey thanks to a series of Soviet ships that were stationed around the globe to receive him as he passed by overhead. On first catching sight of the Earth through his window, he began to describe the small cumulus clouds and the shadows they cast on the Earth's surface. Then he broke from the description and simply exclaimed, 'Beautiful! Beautiful.' Afterwards, on Earth, he would say that the view from 175 to 300 kilometres up was very sharp, similar to that from a high-flying jet. He saw large mountain ranges, large rivers, large forest areas, shorelines and islands.⁹⁰ Throughout the flight, he kept repeating that he felt great.

About eighteen minutes into the flight, the Sun began to set and Gagarin turned off the brighter lights in the capsule to see if he could see the stars. Then, suddenly he was plunged into darkness as the Sun disappeared below the horizon and the stars truly shone. He noted that they looked brighter and clearer than from the Earth's surface. Fifty-seven minutes into the flight, Gagarin saw our planet's atmosphere, hugging the Earth's horizon. He told the ground controllers, 'I can see the Earth's horizon. It has a beautiful blue halo. The sky is black. I can see stars – a pretty fantastic view.'

Twenty minutes or so later, Gagarin's capsule completed its orbit of Earth and began its descent through the atmosphere. As planned, he ejected and parachuted back to Earth. He landed in a field near Saratov, where he had begun his career studying tractors. He was approached by a woman and a child.

'Don't be afraid, comrades!' he called, 'I am a friend.'

'Have you come from space?' the woman asked.⁹¹

NASA sent its first man into space the following month when Alan Shepard finally blasted off on a short sub-orbital flight that lasted fifteen minutes. By the end of that month, President Kennedy had won the backing of Congress to commit the country to sending a man to the Moon by the end of the decade. This was a race that his advisers thought America could win, but even if they didn't win it they thought, 'it is better for us to get there second than not at all … If we fail to accept this challenge it may be interpreted as a lack of national vigor and capacity to respond.'⁹²

In July that year, American Gus Grissom made a similar flight to Alan Shepard's, but it took until the following February for NASA to place John Glenn into space for three orbits of the Earth. By this time the Soviet Union's Gherman Titov had made seventeen orbits in Vostok 2, in a mission lasting 23.5 hours. There could be no doubt, Russia was winning the space race, and their technological success was unacceptable to America's leaders.

President Kennedy faced other international humiliations too. The Cuban Missile Crisis and Vietnam War were fostering social unrest in the country. Traditional authority was seen as failing, and a counterculture movement sprang up to champion a different way of doing things. In an echo of theosophical thinking, it stood for an equal, enlightened society that was engaged with the natural world in a more spiritual way, rather than one that was out to conquer it with technology. So even as Kennedy stood at the Rice Stadium in Houston, Texas, on 12 September 1962 justifying the Moon landings as a demonstration of the inalienable human urge to explore, there was a growing resentment of the time, money and effort that was going into the endeavour. By the middle of the 1960s, around 5 per cent of the entire government budget was being ploughed into NASA, most of it going on the Moon landings.[93] And this was too much for most.

Public opinion polls throughout the 1960s consistently showed that between 45 and 60 per cent of Americans believed that the Apollo Moon landing programme was not worth the money.[94] Yet, paradoxically, apart from the eye-watering cost a clear majority of people supported

NASA's overall goals. This was perfectly summed up in the *Rome News Tribune* for 2 February 1971, which reported that two hundred black protestors had marched to Cape Canaveral (then named Cape Kennedy), Florida, on the morning of the launch of Apollo 14. They were protesting at the billions of dollars being spent on the Moonshots, when many on the march were finding it hard to make ends meet. One of the protest's leaders, Hosea Williams, was quoted as saying, 'We are not protesting America's achievements in outer space, we are protesting our country's inability to choose humane priorities.' And when asked about the spectacle of lift-off itself, he said, 'I thought the launch was beautiful. The most magnificent sight I've seen in my whole life.'

Digging further into the details of opposition to the space programme, historian Matthew Tribbe, of Fullerton College, California, found that opinion about America's lunar landings cut across all demographic groups.[95] There was no simple pattern of support between young and old, male and female, politically right and left, scientists and artists. Being for or against space exploration was a totally new category of its own. It showed how space exploration was something genuinely new in the human experience, which meant that an individual's reaction could not be predicted from a knowledge of their other preferences. It simply did not fit into any historical framework, and therefore could not be judged by past experience.

In the same way that Jung thought we projected our hopes and fears into the night sky, so the space programme became a lightning rod for our inner beliefs. Those who believed in the power of progress saw the achievements of the space race as the herald of a bold new future in which technology would help solve our problems on Earth. Those who didn't share this belief in technology saw space exploration as a tragic betrayal of basic humanity. It had turned the serenity of the night sky into a grubby place of political rivalry, especially since the only way to get there was to use technology that had been developed to rain nuclear oblivion on us all.

Meanwhile stories about space began to turn towards darker themes, where space exploration itself was the villain. In Michael Crichton's novel, *The Andromeda Strain*, published in 1969, a space probe returns to earth having been contaminated with an alien microbe that proves lethal to any human who comes into contact with it. In the classic zombie movie *Night of the Living Dead*, released in 1968, the dead are reanimated by extraterrestrial radioactivity brought back to Earth by a space probe that exploded in our planet's atmosphere.

While there had always been science fiction stories about aliens invading us, such as H. G. Wells's 1898 *The War of the Worlds*, these new stories were different. It was not extraterrestrials but human hubris that brought danger and destruction to Earth. *Star Trek* was a notable exception. Premiering in 1966, it was a throwback to the largely optimistic films of the 1950s that Chesley Bonestell had worked on. It presented a utopian vision of humans and aliens working together to explore the unknown – and initially it bombed. It was cancelled after three years of mediocre ratings in the summer of 1969.[96]

★ ★ ★

While most who opposed Apollo did so because they felt the money could be better spent elsewhere, there were some who voiced a kind of existential fear that travelling into space would sever our connection to the Earth and even rob us of our humanity.

Hannah Arendt was an American philosopher. In 1958, she published her book *The Human Condition*, a discussion of how human activities throughout history have always had unintended repercussions. Like many others, she argued that Galileo's telescopic discoveries (and those of his contemporaries) were a tipping point in history. Arendt defined this as the moment when the accumulation of knowledge passed from philosophers to scientists. This was no trivial matter, she argued, because philosophers derived their knowledge from passive acts of contemplation, whereas scientists did so from actively building equipment, performing experiments, taking measurements. In this handover, she suggested, doing rather than thinking about things became automatically associated with the accumulation of knowledge and thus making progress. In the process, she argued, thinking about why we do things had been lost.

Arendt suggested that as scientists continued to experiment, they made it possible to do things that did not naturally happen on earth. For example, splitting the atom and the subsequent development of the hydrogen bomb unleashed on our planet the same reactions that powered the stars – a process that had formerly only been possible in the depths of space, far from our world. She thought that in bringing these powers down to Earth we were alienating ourselves from what made Earth and earthly living unique. She also felt that the theories developed in the twentieth century that had made these giant leaps possible (quantum theory and relativity) were rendering the universe less understandable because they described reality in such bizarre mathematical fashions that we had no way of truly grasping their meaning. Instead, we simply used these theories' mathematical recipes to blindly make things happen without thinking about whether we really should.

Sputnik-1 was a singular event for her, unlike anything that had yet taken place in history, because our penetration of the night sky meant that we would inevitably send humans into it. So, in her book she proposed that, instead of a headlong rush to the Moon, there should be some reflection and a discussion of what the repercussions of sending humans into space might be. If nothing else, she argued, such a discussion would give us a chance to develop an understanding of why we want to explore space, and find ways of expressing this in a language that everyone can understand.

The reason for Arendt's caution was that she believed space would provide the ultimate 'Archimedean point' from which to view the Earth. This was a scientific concept in which one is so far removed from the object under study that it can be viewed entirely objectively. This god's-eye view of the action is what scientists value most of all, as it allows them to understand something in which they are not involved. But Arendt feared that by jumping into space we would stop seeing the Earth as our home. We would lose our sense of what it is like to experience Earth in all its messy detail, and instead interpret it as nothing but strings of cold data and mathematical relationships.

It was against this backdrop of indifference and dire warnings that the Moon landings took place.

★   ★   ★

The first touchdown happened on 20 July 1969. The mission was called Apollo 11 and the astronauts that reached the lunar surface were Neil Armstrong and Edwin 'Buzz' Aldrin. For that brief moment, the public's objections were largely forgotten and the world united in a way that has seldom been experienced before or since. Half a billion people (15 per cent of the world's population) watched the television coverage of Neil Armstrong becoming the first man to walk on the Moon. It is still the most watched television programme in US history with an estimated domestic audience of 125–150 million people.[97]

In the Soviet Union, the event was downplayed, with only sparse newspaper coverage. The cosmonaut Alexei Leonov, who had performed the world's first spacewalk in 1965 when he spent twelve minutes outside his spacecraft, summed up his feelings as a mixture of envy and admiration. In his book *Two Sides of the Moon*, he remembered hearing of the Apollo 11 launch and thinking that if he himself could not be the first person to walk on the Moon, then he hoped the Americans who had just taken off would make it. In truth, Russia had been effectively out of the race since 1966 when their chief designer, Sergei Korolev, had unexpectedly died on the operating table while undergoing a routine operation for a stomach complaint. With Korolev gone, the Soviet space programme lacked focus amid the competition to succeed him.

And so, in an extraordinary feat of technical and economic accomplishment, America won the most cherished prize of the space race. Upon their return to Earth, Armstrong and Aldrin were national heroes, celebrated wherever they went. But the achievement also sharpened criticism, and resentment hardened into revulsion. Space exploration was increasingly characterised as an élitist, white man's luxury that embodied the inequality that was poisoning American society. Instead of looking at the night sky and feeling awe, many in America now looked up and saw injustice.

To fill the void created by their rejection of a technological future, a growing number of people were rediscovering the kind of spirituality that Blavatsky had promoted through theosophy. As a result, astrological and esoteric beliefs began to proliferate, and the counterculture transformed into the New Age movement.

New Age thinking was characterised by its adoption of the esoteric idea that the physical universe was an illusion hiding the true spiritual reality. This justified opposition to the space programme: if the physical world was viewed as an illusion, then space exploration was the greatest folly of all time. Instead, New Age practitioners could go on their own astral voyages through the use of mind-altering drugs. This would prepare them for the coming spiritual revolution that would lead into the astrological age of Aquarius.

Society as a whole turned its back on space exploration

so completely that just one year after Apollo 11, most Americans struggled to even remember Armstrong's name.[98] Despite the launching of six more Apollo missions, the only one that generated even a fraction of the same public interest was Apollo 13, and that was because an explosion on the spacecraft nearly cost the astronauts their lives. The politicians lost interest too. Having achieved its Cold War goals, Washington slashed NASA's budget, forcing the agency to cancel the last three of its planned ten Moon landings. The final mission took place in December 1972, by which time society had already moved on. Our connection with the night sky and the surrounding universe appeared to be severed once and for all.

But, as with so much about the night sky, looks can be deceiving.

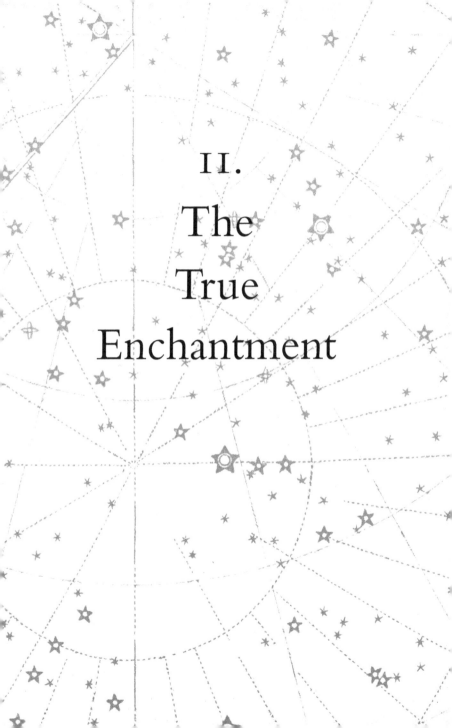

II.

The
True
Enchantment

From the earliest of times humans have looked into the night sky and wondered both what it is, and why it is there. In pursuing an answer, we have always sought to link ourselves in some way to the starry realms. At various times in history we have sought a connection through the action of gods, or through mysterious influences that predict future events, or as a means of moulding our personalities. With the advent of science, it seemed as if this persistent little bit of wish fulfilment could be finally put to rest. But in the middle of the twentieth century something extraordinary happened.

A small number of scientists proved that a fundamental connection did indeed exist between us and the night sky. Their work even implied a reason for why we are surrounded by a giant cosmos. And because of this, they sowed the seeds for a modern enchantment of the universe; one that will endure for as long as there are people to look at the stars. That is because this enchantment derives from testable science rather than some mystical opinion. But those scientists didn't set out to do this at all. Instead, all they wanted to know was what made the stars shine.

By the early decades of the twentieth century, physicists had discovered the true nature of matter. The secret was that everything was composed of particles. The English physicist J. J. Thompson had made the first breakthrough in 1887, when he discovered that the so-called cathode rays, which are emitted from the negatively charged terminal of an electrical circuit, are composed of subatomic particles. He called these particles electrons. New Zealand-born physicist Ernest Rutherford took the next step in 1909 when he showed that atoms consist of a central nucleus surrounded by electrons. The nucleus itself was composed of two different types of particles: protons and neutrons.

The number of protons determines the chemical identity of the atom. For example, hydrogen contains a single proton; oxygen contains eight. In these investigations it became clear that atomic nuclei store energy that can be released when they interact with one another. These newly recognised interactions were called nuclear reactions and followed two basic paths: fusion and fission. In the first, lighter nuclei join together to make a heavier nucleus; in the second, a heavier nucleus splits into two or more lighter nuclei.

In the 1920s, British astrophysicist Arthur Eddington suggested that nuclear reactions could power the stars, and British-born astrophysicist Cecilia Payne used her doctoral thesis to show that the stars were composed of mainly hydrogen and helium gas. This meant that those two elements were by far the most abundant chemicals in the universe. Together, these works inspired a steady succession of researchers who gradually put together our modern understanding of how stars generate their energy.

It is now established beyond doubt that hydrogen and helium make up about 98 per cent of the atomic mass of the universe. The other two per cent is composed of all the other chemical elements combined: the silicon in the rocks, the nitrogen in the air, the oxygen in the water, etc.

The hydrogen and helium were forged during the formation of the universe itself: the moment astronomers refer to as the big bang. Back then, about 14 billion years ago, they were essentially the only elements that existed. In the 1950s, British astrophysicist Fred Hoyle and collaborators authored a series of papers that laid out the theory of how the other two per cent of elements has been built by nuclear reactions inside stars. The basic idea is that at the centre of each star is a natural nuclear reactor, building up elements from hydrogen to iron. These elements are then disgorged into space at the end of the star's life.

Sometimes the stars expel the elements via gentle winds at the end of their lives, as they run out of hydrogen fuel to convert. This will be the fate of our Sun in around 4.5 billion years' time. At the other end of the scale, any star more than five times the mass of the Sun will die in a gigantic explosion known as a supernova. The star literally blows itself to pieces, and in this massive outpouring of energy it simultaneously builds the chemical elements heavier than iron and scatters them through space. Such supernovae were responsible for the 'new stars' that both Tycho and Kepler saw, and modern astronomers have identified the still-expanding clouds of gas and dust produced by these ill-fated stars.

As Hoyle and his colleagues made these tremendous leaps they began to realise the mind-blowing implication of their work: the iron in our blood, the oxygen in our lungs, the carbon in our DNA were all made in the hearts of stars billions of years ago. It took billions of years for the universe to produce the generations of stars needed to produce the elements necessary to make planets like the Earth, and life like us. And during the time it took to do this, the vast cosmos has expanded into the vista we see around us today.

In the library of St John's College, Cambridge, a letter details how Hoyle broke this momentous news to his wife. He told her quite casually that the iron in her kitchen sauce-pans had been made in ancient stars. The young astronomer Carl Sagan was more erudite in 1966 when he wrote that 'our bones are made of calcium formed [in an ageing star] billions of years ago.'[99] But it took Canadian singer-song-writer Joni Mitchell to distil this scientific conclusion into its most memorable epithet in her song 'Woodstock', written in 1969, whose chorus begins, 'We are stardust ...'

With these conclusions, science had provided the most intimate link possible between us and the surrounding universe, and this knowledge formed a bridge between science and New Age thinking by providing a tangible connection to the cosmos. And it did not stop there. Even though there was vociferous dissent over the cost of the Apollo moon landings, it was also undeniable that people were forming an emotional connection to the images returned from the missions. But it was not the images of the lunar surface that had the biggest impact on the general public; it was the views of our own planet – and it had happened by accident in the year leading up to Neil Armstrong's famous small step.

★   ★   ★

Less than a year before America ultimately won the race to the Moon, the Soviets had scored another big victory. In September 1968 they launched Zond 5 on a trip to the Moon and back. The spacecraft carried biological samples including flies, worms, planets, seeds and turtles, and a human mannequin fitted with radiation sensors. It was a clear sign to the Americans that the Soviets intended to send cosmonauts to at least orbit the Moon in the near future, probably before the end of the year. Given the eye-watering amount of money that NASA were spending, they could not be upstaged again.

So when technical hitches meant that the Apollo launch schedule had to be re-jigged, officials at NASA came up with a new plan to send three astronauts into lunar orbit before Christmas. It was unanimously agreed, and on 21 December 1968, Apollo 8 blasted off from Cape Canaveral. Inside were astronauts Frank Borman, Jim Lovell and William Anders. Just three hours and thirty-six minutes into the flight, they reached the furthest humans had ever flown in space, passing the record of 1,369 kilometres set in 1966 by Pete Conrad and Dick Gordon in Gemini 11. As they coasted outwards, they looked back at Earth, and saw a sight no human had ever seen before. The whole Atlantic Ocean was in sight, framed by landmasses: the Americas to the west, Europe and Africa to the east.

'We have a beautiful view of Florida now,' Lovell told mission control, 'and at the same time I can see Africa. West Africa is beautiful. I can also see Gibraltar at the same time I'm looking at Florida.'[100]

Mission control told him to snap a picture of it and then proceeded with the scheduled job of setting up a communications system check. While that was happening, however, mission control could not resist asking what the view was like now. 'I can see the entire Earth now out of the centre window. I can see Florida, Cuba, Central America, the whole northern half of Central America, in fact, all the way down through Argentina and down through Chile,' said Lovell.

Some time in the next ten minutes or so, the crew captured the first image ever taken by a human of the whole Earth. When it was developed in the NASA photographic labs after the mission returned it was stunning: the southern hemisphere was at the top of the image with South America in full view; clouds blanketed much of the globe, but Florida, from where the astronauts had left, was sunny, poking out from behind a thick layer of cloud. The blue Atlantic Ocean dominated the lower half of the globe with the coastline of west Africa just visible to the left of the image, where night is creeping across the continent, giving the Earth a gibbous phase.[101]

But as beautiful as it was, it wasn't the most striking image of the Earth that the crew captured on that mission.

By Christmas Eve 1968, the Apollo 8 command module was on its fourth orbit of the Moon. Anders was photographing the lunar surface so that potential landing sites could be identified for subsequent missions. All was routine, when suddenly he exclaimed, 'Oh my God, look at that picture over there.'

Earth was rising above the Moon's horizon. Anders turned his camera and captured the scene, producing the iconic picture: Earthrise. Although home to around 3.5 billion people at the time, any one of the astronauts could eclipse their home planet by raising nothing more than their hand.

Remembering the moment in 2018, Anders wrote, 'I thought of my wife and five children on that little planet.

The same forces that determined their fates worked on the other three-and-a-half billion inhabitants. From our tiny capsule, it seemed as if the whole Earth was smaller even than the space the three of us inhabited.'[102]

Later that day, the astronauts were making a live television broadcast back to Earth. Their audience was estimated to be around half a billion people, and they showed the world what they had witnessed earlier. Pointing the black-and-white television camera out of the window they beamed back images of a half-illuminated Earth hanging above the rugged lunar landscape. As audiences took in the extraordinary sight, Lovell said, 'The vast loneliness is awe-inspiring and it makes you realise just what you have back there on Earth.'

Shortly afterwards, the astronauts read from the Bible.[103] They had been told during the planning for the flight to say something appropriate. But what words were appropriate when you are an engineer, not a poet, addressing your countryfolk (and many others) from a viewpoint that no one has ever experienced before? The astronauts had failed to reach a decision, so included their wives in the discussion. When that turned up nothing useful, they included friends. Finally one of those came up with the suggestion: read from the Bible. And so they read the first ten verses of the book of Genesis. But it was the view of Earth, not the religious message that truly resonated with viewers.

One of those watching back on Earth was the American poet Archibald MacLeish. He recognised the significance of the view immediately and penned a prescient analysis for the *New York Times*, which was published just hours later on Christmas Day.

Entitled 'Riders on Earth Together, Brothers in Eternal Cold', his piece described how humankind's conception of itself has always depended upon its notion of the Earth, and implicit in his statement is that our notion of Earth has always been determined by our relationship with the night sky.[104] MacLeish illustrates this by saying that Earth was once the centre of the universe, with heaven above and hell below; God was in his heaven and humans were his sole concern. Then came the scientific revolution and the dethronement of God as the divine architect of creation. In this atheistic view of nature, humans were the accidental inhabitants of a physical universe that was neither designed for them, nor set up to ensure their eternal survival. MacLeish actually called us 'the helpless victims of a senseless farce'. And bringing the power of the stars down to Earth in the form of hydrogen bombs put the human race 'beyond the range of reason ... lost in absurdity and war'.

With the circumnavigation of the Moon and the new view of Earth that this afforded us, MacLeish hoped that we had reached another watershed; that seeing Earth embedded in the vast black of the night sky would cause us to reappraise what it means to be human. Having swung from thinking of ourselves as the chosen race of God, to the victims of senseless technological development, he now thought that we had harnessed that technology to truly look at our place in the universe. And with this new view, MacLeish hoped that 'man may at last become himself'. Now we could at last realise that the future is in our hands, that we make our destiny by the choices we take.

He also hoped that this new realisation would bring the human race together in a way that had eluded us before. He wrote: 'To see the earth as it truly is, small and blue and beautiful in that eternal silence where it floats, is to see ourselves as riders on the earth together, brothers on that bright loveliness in the eternal cold – brothers who know now they are truly brothers.'

The full colour Earthrise image inspired awe not just in America but around the world. Its power lay in that it offered a view of ourselves that we had never seen before. Throughout history we had looked up into the darkness and seen the celestial objects framed by the black of the night sky, but we had never seen Earth in this way. Now that we had this viewpoint, it was visceral confirmation that we were an integral part of the surrounding universe – not somehow removed from it. We knew this intellectually before, but seeing it with our own eyes transformed that knowledge into an emotional experience, and for most people, that was the difference needed to drive home this sense of connection to the planet and the wider cosmos. It was the exact opposite of what Hannah Arendt had feared might happen when humans looked back at Earth. Thankfully, instead of becoming detached from our humanity, we found it had been reinforced.

In 1983, author Don DeLillo summed up the same feelings in his short story, *Human Moments in World War III*. Right at the beginning of the story, an astronaut is looking at the Earth and DeLillo describes how 'The view is endlessly fulfilling. It is like the answer to a lifetime of questions and vague cravings.' And the importance of the Earthrise image has only increased with time. In 2003, the editors of *Life* magazine published the *100 Photographs that Changed the World*. The main image they chose for the front cover was Earthrise. In the commentary, American wilderness photographer Galen Rowell described it as 'the most influential environmental photograph ever taken'. Indeed it symbolised the West's nascent green movement perfectly. One Earth, one humanity, one home.

Anders summed it up perfectly in 2018 on the fiftieth anniversary of his taking the image. He wrote, 'We set out to explore the Moon and instead discovered the Earth.'[105]

On the final Moonshot, Apollo 17, it was the image they took of Earth just over five hours into the flight that made the biggest impact.[106] Taken from a distance of about 45,000 kilometres, it captured the whole illuminated globe of the planet.[107] Africa is squarely in the centre of the image, with the golden Sahara desert clearly visible. Being December, it is summer in the southern hemisphere and the fully illuminated Antarctica ice cap dominates the image. White clouds curl across the globe and in the top right of the image is the Tamil Nadu cyclone that had killed eighty people. It perfectly encapsulated both the beauty of the planet and the fragility of life on Earth.

The image was released to the media on Saturday 23 December 1972, and again it created an immediate sensation, making it to the front page of almost every newspaper on the planet. Now referred to as the blue marble shot, it is easily the most important thing to come out of Apollo 17. It was released into the public domain and has become one of the most – if not the most – reproduced images of all time. It too has become a symbol of the environmental movement.

★   ★   ★

As the astronauts returned to Earth, another wholly unexpected outcome of the space programme began to manifest itself whenever they tried to put their experiences into words. In their training they had learned exactly how their spacecraft flew, about why the trajectories worked and everything that was scheduled to happen during the mission, but nothing had prepared them for the emotional effects that being in space would produce. As they spoke of their experiences, it was clear that all of them felt changed at a deeply fundamental level.

Gene Cernan, who walked on the Moon as part of the Apollo 17 mission, told *The Atlantic* magazine, 'You have to literally just pinch yourself and ask yourself the question, silently: Do you know where you are at this point in time and space, and in reality and in existence, when you can look out the window and you're looking at the most beautiful star in the heavens – the most beautiful because it's the one we understand and we know, it's home, it's people, family, love, life – and besides that it is beautiful. You can see from pole to pole and across oceans and continents and you can watch it turn and there's no strings holding it up, and it's moving in a blackness that is almost beyond conception.'[108]

Many astronauts spoke of feeling a profound connection to the cosmos and to perceiving the unity of all life on Earth – views that would not have sounded out of place coming from some New Age guru who aspired to developing a 'higher consciousness'.

On a more human level, Russell 'Rusty' Schweickart flew into orbit around the Earth as part of the Apollo 9 mission. His wife jokes that he went into space in love with her, but came back in love with the planet Earth. His epiphany happened in the middle of a spacewalk. As usual for spaceflight, the activity had been scheduled to the minute, but when a camera broke down, instead of working through the constant stream of instructions from mission control, Schweickart found himself hanging above the Earth with time to reflect. The question that came into his mind was how a kid from New Jersey got to be in his position – not so much being in orbit but being in a way a sensory envoy for the human race. 'I kind of declared myself to be a sensing element ... like an eyeball in a set of ears and eyes out on the end of a finger ... as humanity begins to move out of the womb ... of Mother Earth.'[109]

Early inklings of the psychological shift that spaceflight can bring can be seen in the way the first astronauts described their views from orbit. John Glenn was the first American to orbit the Earth. For the rest of his life, when asked what it felt like to be the first American in orbit, his answers often steered clear of national pride. Instead he spoke of the great beauty of our planet. In particular, he remembered the sunrises, which came every ninety minutes or so as he raced around the Earth. In 1997, he told *American History Magazine* that when you're on Earth, sunrises are golden and orange, but from space all the colours of the rainbow take part in the breathtaking experience due to optical effects in the Earth's atmosphere.[110]

For the astronauts who spent days in space, and especially those who had gone to the Moon, the feelings were magnified. Edgar Mitchell, who walked on the Moon as part of the Apollo 14 mission, was very aware of the change in his perspective. After several years of reflection, he told *People* magazine in 1974, 'You develop an instant global consciousness, a people orientation, an intense dissatisfaction with the state of the world, and a compulsion to do something about it. From out there on the Moon, international politics look so petty. You want to grab a politician by the scruff of the neck and drag him a quarter of a million miles out and say, "Look at that, you son of a bitch".'

He also found that previously known scientific facts suddenly took on whole new levels of meaning. He could see ten times more stars from orbit than from Earth, and this started him thinking about the recent scientific revelation that most of the atoms in his body and in his space capsule were built in the hearts of giant stars billions of years ago. He was struck by an ecstatic feeling of connection to those stars. 'I realised that the molecules of my body and the spacecraft and my companions were prototyped in an ancient generation of stars. And somehow it suddenly became very personal, instead of an objective, "Oh, yes. Molecules and atoms were made in those stars." No. *My* molecules were made in those stars, and this was a "wow!"'[111]

When he got back to Earth, he began searching for an explanation for his feelings, or at least a similar description, in order to know that someone else had experienced this extraordinary state of mind, but he found nothing that could help in the realms of the science he was familiar with, and nothing either in the religious texts he looked at. Then an ancient document called *Yoga Sūtras of Patañjali* was brought to his attention. It collected the wisdom and practice of yoga. Part of that was a description of the meditative state of *savikalpasamadhi*, in which the individual concentrates on an object so fully that their conscious state transforms into one of blissful unity with their surroundings. And in this description, Mitchell recognised what had happened to him. The most extraordinary part was that the *Yoga Sūtras* had been written more than 1500 years before Mitchell flew into space.

Everyone coming back from those early missions was changed in a way that could not be undone when the experience was over. To have seen the Earth from above drove home the fact that our world is a single beautiful entity, and its people a united whole (even if they believed themselves to be a series of divided nations). Mitchell summed it up in a single line: 'We went to the moon as technicians. We returned as humanitarians.'[112]

Apollo 11 astronaut Michael Collins, who piloted the lunar command module while Armstrong and Aldrin walked on the Moon, found the experience so positive that he said, 'The pity of it is that so far the view [of Earth from space] has been the exclusive property of a handful of test pilots, rather than the world leaders who need this new perspective, or the poets who might communicate it to them.'[113]

Although the Moon landings came to an end, America and Russia continued to send people into orbit, and the more who saw our planet from above, the more evidence accumulated that the experience produced a real cognitive shift. In 1987, using the testimony of twenty-five astronauts, author Frank White published a book on the subject and for its title coined the term *The Overview Effect*. He described the common threads of this effect as being a sense of awe at the beauty and fragility of the planet, which leads to an understanding of the interconnectedness of all life and a renewed sense of responsibility for taking care of the environment.

In 2006, American psychiatrist Eva Ihle sent an anonymous questionnaire to 175 astronauts and cosmonauts, asking about their experience of being in space. Of the thirty-nine who replied, each one said that being in space was a meaningful experience that led to them making positive and enduring changes in their attitude and behaviour. The strongest of those reactions was always about the beauty and fragility of Earth. Interestingly, Ihle found that the response did not vary by demographic group, nor the number of missions flown, nor the total time spent in space.[114] In other words, once you had seen the Earth from this god-like vantage point, you were changed by it for ever.

In recent years, a number of studies have found that the emotion of awe – so clearly present in the overview effect – can have a transformative effect on our lives. It makes people feel more altruistic, less stressed, less time-pressured and less materialistic.[115] It is as if awe diminishes people's sense of self-importance and makes them more willing to behave in a way that's good for the collective. The overview effect experienced by astronauts is simply an extreme example of this – and, as the *Yoga Sūtras* proved, not the only way of achieving it.

Watching a natural phenomenon such as a sunset; gazing over a huge landscape; meditation; studying a great work of art or experiencing a musical performance; all of these can also generate the same feeling of awe. So too can understanding something from a new point of view, perhaps understanding the scientific explanation for an otherwise mysterious phenomenon. As stated at the very beginning of the book, the easiest way to experience awe is by standing beneath the night sky and contemplating the stars. Yet this is the era when the light pollution of urban living has largely extinguished the stars from our view, and because of this, for many people the easiest way to generate awe and feel the connection to the cosmos that our ancestors naturally felt has been completely lost. Or has it?

★　★　★

The truth is that the universe is at our fingertips as never before. For the first time in history, one does not need to be beneath the night to feel connected to the cosmos. The space programme has now sent robots to all corners of our Solar System. Space telescopes peer into the depths of space further than at any time in history, and instead of those images being stored in university campuses, they are freely available over the Internet. Now almost any individual can look through a robot's eyes as it trundles the deserts of Mars, or witness extraordinary solar storms which explode with the energy of a billion atomic bombs. We can see distant galaxies caught in the throes of colliding with each other, or peer into stellar nurseries where stars are being born. As of 2019, we can even see the silhouette of gigantic black holes that lurk in the centre of nearby galaxies.

All this wonder is available at the touch of a button; it is the ultimate on-demand reality show, and it is fostering a relationship with the night sky that is unique in history. We now see the universe in full colour, in close-up, and tangible in a way we have never experienced before; and because of this, the night sky is closer to us than ever before. Accurate portrayals of deep space and planets can now be seen on television and in films, both in factual documentaries and fictional television series. The first thing you see in George Lucas's epochal 1977 film *Star Wars*, just before the spacecraft come thundering over the top of the screen, is the beautiful blue atmosphere hugging the fictional planet Tatooine – as Yuri Gagarin described Earth's atmosphere on his spaceflight.

At the vanguard of this new cosmic perspective is the Hubble Space Telescope. Since its launch in 1990, it has been providing a steady stream of remarkable views that have gone a long way to reconnect us with the universe. They have shown us the furthest reaches of deep space, exploding stars, cannibal galaxies, collisions between a comet and a planet – the list of wonders goes on and on. Even if you don't understand the physics of what you are looking at, the images themselves are aesthetically beautiful, and because they represent something so much larger than us they can generate awe. Gradually, these images and the others produced by the myriad probes now in space are reconnecting us to the universe.

My first glimpse of this new connection was during the 6 August 2012 landing of NASA's Curiosity rover on Mars. I was up early to follow the landing over the Internet, and I was struck by the sheer number of people talking about it on social media. I was also struck by the crowds that had gathered in Times Square, New York, to watch the live-feed of the event. But perhaps what was most interesting of all was that there were no cameras on the descent module. We were watching the faces and the reactions of the NASA mission controllers, vicariously experiencing their connection with the spacecraft, their connection with the universe.

I believe that we are experiencing a re-enchantment of the night sky that is not based on mystical associations or occultism. This true enchantment comes from the science of the night sky, combined with the images that technology has made it possible for us to capture. The combination of the beautiful images and mind-blowing facts generates awe, both at the universe itself and at the human achievement of unlocking its secrets.

By reaching into the night sky, we can now look back at Earth and feel awe-struck by its beauty. We have inverted the traditional way of thinking. It is no longer space that is the awesome thing: it is the fragile balance of Earth. This is the true legacy of the space race – not Teflon or any of the other supposed technological spin-offs, but the realisation that our home planet is not as big as we think from the surface.

The wider truth is that the night sky now has a greater influence on us than at any other time in history. We are more connected to the cosmos than ever before, yet unaware of it because we have normalised it. We now use satellites for communications, for weather forecasting, for navigation. There are even prototype clocks that gather the radio signals from fast-spinning stars to tell the time more accurately than most clocks on Earth.[116] We use space now for all the same things that the earliest hunter-gatherers used it for. We have come full circle – we just do it more reliably nowadays because of technology.

And despite the preponderance of city living, people are seeking out the night sky. It has been widely reported that the travel industry saw a rise in so-called 'astrotourism' during 2017.[117] These are trips to see eclipses, the northern lights or simply to experience the feeling of being beneath a really dark night sky. And as companies develop and sell space tourism, the number of people experiencing the overview effect by seeing our planet from space is only set to increase. More and more of us, it seems, are destined to feel what Fred Hoyle, the astronomer who discovered that we are made of star dust, knew intellectually when he said this: 'Space isn't remote at all. It's only an hour's drive away if your car could go straight upwards.'[118]

I truly experienced an epiphany in 2005 when I visited the European Southern Observatory's aptly named Very Large Telescope (VLT) atop Cerro Paranal in Chile.

I was standing with the other astronomers on the mountain top at dusk. Surrounded by the red earth of Chile's Atacama desert, and dwarfed by the shining metal buildings that house the VLT's four giant telescopes, we stood with our backs to the Sun and watched as the colour drained from the sky. We were waiting for the Sun to touch the horizon because in the few minutes it would take to disappear completely, we would see the silhouette of Earth reach up into the sky, and in that creeping darkness, we would look for ourselves.

As the spectacle began, those around me stirred. First, we saw the dark outline of the mountain rising up, projected onto the sky. Next, we saw the telescope buildings appear as square blocks on the mountain top. Now came the true test.

If the air were particularly still, then just before the sunlight disappeared completely, we would see our own silhouettes appear on the night sky. The calmer the air, the easier it would be to see our silhouettes and the better that night's view of the universe would be.

My heart quickened. I thought that I did indeed see little shapes that resembled human forms. Could I have been that lucky? I glanced at my companion, who smiled.

'It's going to be a good night,' he said.

I looked straight back at the sky but already the view had disappeared to be replaced, one by one, by the stars. Soon they carpeted the sky in a way that would seem impossible for city dwellers to believe. There were so many that they quickly buried the familiar constellations – the signposts I use to find my way around the night sky. It was a dizzying sensation, to be lost in something I thought I knew so well. And it was then that I felt an overwhelming feeling that resolved into a kind of euphoria.

Around me the telescopes started to work. Guided by computers, their buildings rotated almost silently and the instruments moved, acquiring their first targets of the night. With the desert earth beneath my feet and the stars above my head; with the cold air on my face and the fire of imagination in my mind, I was beneath the cloak of night yet connected to the cosmos. In that moment, my longing to know the secrets of the night sky burned more brightly than ever. I felt a sense of profound calm accompanying my euphoria, a definite feeling of privilege to have witnessed a moment of such extreme natural beauty, and I felt part of the greater universe.

For that brief moment, I was not an autonomous living thing, navigating my way through the external world; I was at one with it all. That stunning view of the cosmos stimulated me in such a profound way that it forced me to confront the fact that I am nothing more than a minuscule part of something much bigger. Although I rely on science and its methods to connect me to the universe, and to satisfy the urge to understand the prime mystery of why it's all there, I now have absolutely no doubt why people born in earlier times or with different cultural values would have automatically imagined that some sort of magical design must lie behind the phenomenon. And if there is design, there is purpose and meaning.

It must have been inescapable feelings such as these that drove the hunter-gatherer aggrandisers to establish their secret societies to investigate this mysterious aspect of the night sky. And in those groupings and their rituals we see the beginnings of religion. And it is in the creation myths that sprang up in the centuries before writing that we see the first attempts at what would become philosophy, astrology, and science.

And with these thoughts, not only did that extraordinary moment in Chile instil in me a deep feeling of connection to the cosmos, but it also generated a deep empathy. Because as long as there are humans in this universe, I know that they will look at the night sky and feel the same way I did in that moment; the same way our ancestors did.

Our connection to the night sky is inescapable, it is instinctive, it is what it means to be human.

Notes

1 A. Marshack, *The Roots of Civilization: the Cognitive Beginning of Man's First Art, Symbol and Notation* (New York: McGraw-Hill, 1972).

2 This translation comes courtesy of Carl Sagan in *Cosmos*, 1980.

3 P. A. Mellars, K. Boyle, O. Bar-Yosef & C. Stringer (ed.), *Rethinking the Human Revolution: new behavioural and biological perspectives on the origin and dispersal of modern humans* (Cambridge: McDonald Institute for Archaeological Research, 2007).

4 In Yuval Noah Harari's bestselling book, *Sapiens*, he refers to this event as the cognitive revolution.

5 The Ishango bone is on public display at the Royal Belgian Institute of Natural Sciences in Brussels.

6 Jean de Heinzelin, 'Ishango', *Scientific American*, 1962, 206:6, pp. 105–16.

7 Richard L. Currier, *Unbound: How Eight Technologies Made Us Human, Transformed Society, and Brought our World to the Brink* (Arcade, 2015).

8 www.newscientist.com/article/
dn24090-how-many-uncontacted-tribes-are-left-in-the-world/

9 Thomas Forsyth McIlwraith, *The Bella Coola Indians*, Vols. 1 & 2 (University of Toronto Press, 1948).

10 B. Hayden & S. Villeneuve, 'Astronomy in the Upper Paleolithic?', *Cambridge Archaeological Journal*, 2011, 21(3), pp. 331–55. doi:10.1017/S0959774311000400.

11 www.archeociel.com/index.html

12 Hayden & Villeneuve, 'Astronomy in the Upper Paleolithic?'

13 J. McK Malville et al., 'Astronomy of Nabta Playa', *African Sky*, 2007, Vol 11, pp. 2–7.

14 Fred Wendorf & Romuald Schild, 'Late Neolithic megalithic structures at Nabta Playa (Sahara), southwestern Egypt', Comparative Archaeology Web (26 November 2000).

15 Amanda Chadburn, 'Stonehenge World Heritage Site, United Kingdom', *ICOMOS–IAU Thematic Study on Astronomical Heritage*, pp. 36–40.

16 Gerald Hawkins, 'Stonehenge Decoded', *Nature*, 1963, 200, pp. 306–8.

17 Gerald Hawkins, *Stonehenge Decoded* (Doubleday, 1965).

18 Jacquetta Hawkes, 'God in the Machine', *Antiquity*, 1967, Vol. 11 (163), pp. 174–80.

19 Mike Parker Pearson, 'Researching Stonehenge: Theories Past and Present', *Archaeology International*, 2013, Vol. 16, pp. 72–83.

20 www.ancient-origins.net/artifacts-ancient-writings/kesh-temple-hymn-5600-year-old-sumerian-hymn-praises-enlil-ruler-gods-021152

21 www.nature.com/articles/35042510

22 Maat's scales continue to be the symbol of justice even today, and are usually depicted with a woman holding them.

23 Ara Norenzayan, *Big Gods: How Religion Transformed Cooperation and Conflict* (Princeton University Press, 2013).

24 Vere Gordon Childe, *Man Makes Himself* (Watts, 1936).

25 https://arxiv.org/pdf/1307.8397.pdf

26 Andrew Curry, 'Göbekli Tepe: The World's First Temple?', *Smithsonian Magazine*, November 2008.

27 Jean-Pierre Bocquet-Appel, 'When the World's Population Took Off: The Springboard of the Neolithic Demographic Transition', *Science*, Vol. 333, 29 July 2011, www.sciencemag.org

28 Norman Lockyer, *The Dawn of Astronomy* (Cassell and Company, 1894).

29 Jay B. Holberg, *Sirius: Brightest Diamond in the Night Sky* (Springer Praxis Books, 2007).

30 www.archaeology.org/issues/99-1307/artifact/935-egypt-limestone-sundial-valley-kings

31 W. Dodd, 'Exploring the Astronomy of Ancient Egypt with Simulations II: Sirius and the Decans', *Journal of the Royal Astronomical Society of Canada*, Vol. 99, No. 2, p. 65.

32 O. Neugebauer, 'The Egyptian "decans"', *Vistas in Astronomy*, Vol. 1, 1955, pp. 47–51.

33 Alessandro Berio, 'The Celestial River: Identifying the Ancient Egyptian Constellations', *Sino-Platonic Papers*, No. 253, December 2014.

34 Jed Z. Buchwald, 'Egyptian Stars under Paris Skies', *Engineering and Science*, No. 4, 2003, p. 20

35 M. W. Ovenden, 'The origin of the constellations', *Philosophical Journal*, 3 (1), 1966, pp. 1–18.

36 Mary Blomberg and Göran Henriksson, 'Evidence for the Minoan origins of stellar navigation in the Aegean', Actes de la Vème conférence de la SEAC, Gdańsk de la SEAC, Gdansk, 5–8 September 1997. Światowit supplement series H: Anthropology II. A. Le Beuf and M. Ziólkowski (eds), 1999, pp. 69–81.

37 B. E. Schaefer, 'The latitude and epoch for the formation of the southern

Greek constellations', *Journal for the History of Astronomy* (ISSN 0021-8286), Vol. 33, Part 4, No. 113, pp. 313–50 (2002).

38 The introduction of scientific observation to our understanding of the natural world was not universally embraced. In the seventeenth century when natural philosophers such as Robert Hooke were beginning their investigations into air pressure, they were accused of utter folly at 'weighing the air'. Yet this measurement went on to be the basis of meteorology, which has saved countless lives with its forecasts and warnings.

39 B. Van der Waerden, 'Babylonian Astronomy. III. The Earliest Astronomical Computations', *Journal of Near Eastern Studies*, 1951, 10(1), pp. 20–34. http://www.jstor.org/stable/542419

40 John Steele, 'Astronomy and culture in Late Babylonian Uruk', 2011 Proceedings IAU Symposium No. 278, 2011, 'Oxford IX' International Symposium on Archaeoastronomy, Clive Ruggles (ed.).

41 Michael Gagarin in *The Oxford Encyclopedia of Ancient Greece and Rome*, Vol. 7, p. 64.

42 Although we still agree with this basic principle today, we differ on the nature of that fundamental matter. Today particle accelerators have shown that fundamental particles such as quarks and electrons go together like subatomic building bricks to form the atoms from which all things are made.

43 Markham J. Geller, *Melothesia in Babylonia: medicine, magic, and astrology in the ancient near east* (Boston: De Gruyter, 2014).

44 http://blog.wellcomelibrary.org/2014/01/the-enigma-of-the-medieval-almanac/

45 Bernard Capp, *Astrology and the Popular Press: English Almanacs 1500–1800* (London: Faber and Faber, 2008).

46 John W. Livingston, 'Ibn Qayyim al-Jawziyyah: A Fourteenth Century Defense against Astrological Divination and Alchemical Transmutation', *Journal of the American Oriental Society*, Vol. 91, No. 1 (January–March 1971), pp. 96–103.

47 If you have access to a piano or keyboard, try it for yourself. The notes in Nicomachus's planetary scale are D, C, Bb, A, G, F, E.

48 Modern astronomers believe that a substance called dark matter pervades the universe. It's an invisible kind of matter, completely unlike the normal atoms that make up stars, planets, you and me. Hence, it can't be seen with normal telescopes. Their hypothesis is that the gravity from dark matter helps to hold together the largest collections of stars, called galaxies. There are dozens – possibly hundreds – of proposals for what these invisible particles could be. Billions of pounds have been spent on highly sensitive

equipment to try to capture it – or even make the stuff. Yet despite decades of effort, no one has been able to detect a single, solitary piece of it. This failure, however, has not dented faith in the idea. The majority of astronomers and particle physicists still believe that dark matter exists in some form or another. To me, the same process is taking place in our modern evaluation of the different possible dark matters and the ancient philosophical discussions of the celestial harmony. Modern scientists and the ancient philosophers alike are simply extrapolating from the knowledge of their day, hoping that they have struck upon the right idea.

49 Others thought there was nothing religious about it and that such movement was a natural consequence of matter. Aristotle, for example, proposed that circular motion was the natural state of movement for anything made from aether. In contrast, he claimed that straight line movement was the most natural form of motion for things made of earth and water, which neatly explained away why things fell off tables straight to the ground but the Moon and other celestial objects followed orbits.

50 Alessandro Bausani, 'Cosmology and Religion in Islam', *Scientia/Rivista di Scienza*, 1973, Vol. 108 (67), p. 762.

51 The Gregorian calendar is not a perfect system but it is a vast improvement. It will now take until the sixth millennium before the calendar has moved a whole day out of step with the equinox.

52 Owen Gingerich, *The Book Nobody Read* (Walker & Company, 2004).

53 Donald V. Etz, 'Conjunctions of Jupiter and Saturn', *Journal of the Royal Astronomical Society of Canada*, 94, pp. 174–8, 2000, Aug./Oct.

54 Margaret Aston, 'The Fiery Trigon Conjunction: An Elizabethan Astrological Prediction', *1 Isis*, Vol. 61, No. 2 (Summer 1970), pp. 158–87, The University of Chicago Press on behalf of The History of Science Society.

55 The pyramids of Egypt are not perfect solids because their bases are squares rather than another equilateral triangle.

56 www2.hao.ucar.edu/Education/FamousSolarPhysicists/ tycho-brahes-observations-instruments

57 The remains of these two exploding stars have since been found and studied by modern astronomers. An astronomical curiosity is that on average we should expect one such supernova to be visible in our galaxy every century. Yet none has been visible since Kepler's. This is thought to be just a statistical fluke rather than a meaningful change in the behaviour of stars in our galaxy.

58 For completeness, it is essential to mention that British astronomer Thomas Harriot sketched the Moon through a telescope on 26 July 1609. Thus, his work predates Galileo's by a few months.

59 E. A. Whitaker, 'Galileo's Lunar Observations and the Dating of the Composition of Sidereus Nuncius', *Journal for the History of Astronomy*, Vol. 9, p. 155.

60 Peter Harrison, *The Fall of Man and the Foundations of Science* (Cambridge University Press, 2009).

61 Instauration is an archaic word which means restoration after a period of neglect or decay.

62 Sorry.

63 Allan Chapman, 'Edmond Halley's Use of Historical Evidence in the Advancement of Science', 1994, *Notes and Records of the Royal Society of London*, Vol. 48, No. 2, pp. 167–91.

64 Newton himself also rejected the idea of the Trinity but had the good sense to keep his belief private.

65 U. B. Marvin, 'The meteorite of Ensisheim – 1492 to 1992', *Meteoritics* (ISSN 0026-1114), Vol. 27, March 1992, pp. 28–72.

66 Hitoshi Yamaoka, 'The quinquennial grand shrine festival with the Nogata meteorite', *Highlights of Astronomy*, Vol. 16, XXVIIIth IAU General Assembly, August 2012 (c) International Astronomical Union 2015 T. Montmerle (ed.) https://www.cambridge.org/core/services/aop-cambridge-core/content/view/S1743921314005225

67 Not all shooting stars result in a meteorite making it to the ground, mostly the shooting star phenomenon is caused by pieces no larger than sand grains burning up in the atmosphere. When the object is larger, however, the bright streak can be much more powerful. It can be visible in the daylight sky and create sonic booms because it is travelling faster than the speed of sound.

68 John G. Burke, *Cosmic Debris: Meteorites in History* (Berkeley and Los Angeles: University of California Press, 1986).

69 N. V. Vasiliev, A. F. Kovalevsky, S. A. Razin & L. E. Epiktetova, eyewitness accounts of Tunguska (crash), 1981.

70 Richard Jenkins, 'Disenchantment, Enchantment and Re-Enchantment: Max Weber at the Millennium', *Max Weber Studies*, Vol. 1, No. 1 (November 2000), pp. 11–32.

71 Foreword to the *Philosophiæ Naturalis Principia Mathematica*

72 www.ebyte.it/logcabin/belletryen/IsaacNewton_OdeByHalley.html

73 Joseph Addison, *The Spectator*, No. 420, 2 July 1712.

74 There is interesting echo of Pythagorean thinking here. Back in the sixth century, Pythagoras defined his concepts of the limited and unlimited, and thought that the music of the spheres could unite the two.

75 J. V. Golinski, 'Sublime Astronomy: The Eidouranion of Adam Walker and His Sons', *Huntington Library Quarterly*, 80:1 (2017), pp. 135–57.

76 The word popularisation was itself coined in the nineteenth century in France. Bernadette Bensaude-Vincent, Liz Libbrecht, 'A public for science. The rapid growth of popularization in nineteenth century France', in Réseaux: *The French journal of communication*, Vol. 3, No. 1, 1995, pp. 75–92.

77 Anna Henchman, *The Starry Sky Within: Astronomy and the Reach of the Mind in Victorian Literature* (Oxford University Press, 2014).

78 www.vangoghletters.org/vg/letters/let691/letter.html#translation

79 https://theconversation.com/flying-chariots-and-exotic-birds-how-17th-century-dreamers-planned-to-reach-the-moon-84850

80 www.translate.google.co.uk

81 Of course, we now know that the Moon is so small that its gravity is too weak to hold an atmosphere at all.

82 www.popularmechanics.com/space/moon-mars/a28485/russian-rocket-genius-konstantin-tsiolkovsky/

83 Final Report, 1952 Summer Study Group, February 10, 1953 (2 Vols.), LLAB.

84 Paul Dickson, *A Dictionary of the Space Age* (The Johns Hopkins University Press, 2009).

85 Sputnik-1 re-entered Earth's atmosphere and burnt up on 4 January 1958 after completing 1440 orbits.

86 Roger D. Launius, 'It All Started with Sputnik: An eminent space historian looks back on the first 50 years of space exploration', *Air & Space Magazine*, July 2007.

87 One hundred and one years later, the Society for Psychical Research issued an assessment of their original report that concluded it was too hasty in its condemnation of Blavatsky. On its website, the SPR says: 'Today, the SPR continues to promote and support the main areas of psychical research, carrying out field investigations, surveys and experimental work. It holds no corporate view about the true origin and meaning of psi – as telepathic and other psychical phenomena are now collectively termed – and debate among its members with regard to particular subjects is often vigorous. However, it's fair to say that from the earliest times the consensus view of its members – and of the psi research community in general – has been that psi is real, and that while the phenomena should certainly be explained in scientific terms, such a science does not at present exist.'

88 Nicholas Campion, *A History of Western Astrology Volume II, The Medieval and Modern Worlds* (Continuum, 2009).

89 Carl Jung, *Letters*, Vol. II, pp. 463–4.

90 www.theatlantic.com/technology/archive/2011/04/
    yuri-gagarins-first-speech-about-his-flight-into-space/237134/

91 www.newscientist.com/article/
    mg21028075-600-yuri-gagarin-108-minutes-in-space/

92 Memo 'Recommendations for our National Space Program: Changes,
    Policies, Goals' from James E. Webb and Robert McNamara to Vice Presi-
    dent Lyndon B. Johnson, 8 May 1961.

93 NASA's budget since the late 1970s has been less than 1 per cent of the
    federal budget.

94 Roger D. Launius, Public opinion polls and perceptions of US human
    spaceflight', *Space Policy* 19 (2003), pp. 163–75.

95 Matthew D. Tribbe, *No Requiem for the Space Age* (OUP, 2014)

96 *Star Trek* eventually became a huge success, but only after a decade of
    re-runs as society forgot about the cost of the Moon landings and simply
    remembered the achievement.

97 www.hollywoodreporter.com/news/neil-armstrongs-moonwalk-killed-
    box-off ice-1969-1149903

98 Tribbe, pp. 8–9.

99 Shklovsky and Sagan, *Intelligent Life in the Universe* (Holden-Day Inc., 1966).

100 https://web.archive.org/web/20080923012425/http://history.nasa.gov/
    apo8fj/03day1_green_sep.htm

101 NASA reference number AS08-16-2593.

102 www.space.com/42848-earthrise-photo-apollo-8-legacy-bill-anders.html

103 Frank Borman interview, Boffin Media, private communication.

104 http://cecelia.physics.indiana.edu/life/moon/Apollo8/122568sci-nasa-ma-
    cleish.html

105 www.space.com/42848-earthrise-photo-apollo-8-legacy-bill-anders.html

106 Catalogue number: AS17-148-22727.

107 www.theatlantic.com/technology/archive/2011/04/
    the-blue-marble-shot-our-first-complete-photograph-of-earth/237167/

108 www.theatlantic.com/technology/archive/2011/04/
    the-blue-marble-shot-our-first-complete-photograph-of-earth/237167/

109 Unpublished interview transcript by Richard Hollingham, Boffin media,
    for *Message from the Moon*, Radio 3, (https://www.bbc.co.uk/programmes/
    m0001psz). Transcript supplied by private communication.

110 'John Glenn: First American to Orbit the Earth', *American History*,
    October 1997.

111 www.theatlantic.com/technology/archive/2016/02/edgar-mitchell/461913/

112 https://www.theatlantic.com/technology/archive/2016/02/
    edgar-mitchell/461913/

113 Michael Collins, *Carrying the Fire* (Farrar, Straus and Giroux, 2019).

114 Eva C. Ihle, Jennifer B. Ritsher, Nick Kanas, 'Positive Psychological Outcomes of Spaceflight: An Empirical Study', *Aviation, Space, and Environmental Medicine*, Vol. 77, No. 2, February 2006, pp. 93–101(9), Aerospace Medical Association.

115 Summer Allen, *The Science of Awe*, September 2018, A white paper prepared for the John Templeton Foundation by the Greater Good Science Center at UC Berkeley.

116 www.esa.int/Applications/Navigation/ESA_sets_clock_by_distant_spinning_stars

117 www.telegraph.co.uk/travel/comment/astrotourism-new-sustainable-travel-trend/

118 Fred Hoyle, *The Observer*, 9 September 1979, 'Sayings of the Week'.

Index

Abd al-Rahman al-Sufi, *Book of Fixed Stars* (*kitabsuwar al-kawakib*), 95
Abu Sa'id al-Sijzi, 128–9
Addison, Joseph, 180–1
aether, 115, 276n49
African Cattle Complex, 26
after-life: Ancient Egyptian, 39, 42, 43–5, 60, 62; Ancient Greek, 110–11; immortality and resurrection of the dead, 202–3; as motivation for co-operation, 45; *see also* heaven
Age of Enlightenment *see* Enlightenment
'aggrandizers,' 18, 36
agriculture: and interest in night sky, 53; neolithic revolution, 34–5, 45–6, 53
Akitu (Babylonian festival), 82–3
Aldrin, Edwin 'Buzz,' 245, 246
alignment of buildings and monuments: palace of Knossos, 74; pyramids, 40–1; sanctuary of Petsophas, 74; stone circles, 27, 28, 30, 32–4, 47–8; temples of Isis, 59–60
almanacs, astrological, 97, 98–9
American space programme: animals in space, 236–8, *238*; Curiosity Mars rover, 267–8; establishment of NASA, 219; Kennedy's 1962 speech, 5, 6, 240; manned space flights, 236, 239–41, 245–7, 255–7, 260, 261, 262, 263–4; Moon missions, 239–41, 245–7, 255–7, 260, 261, 262, 263–4; satellites, 210–11, 218–19
Anaximander, 90
Anaximenes, 90
Ancient Egypt: after-life, 39, 42, 43–5, 60, 62; calendars, 58–60, 61; gods, 38–9,

44–5, 59–60; hieroglyphs, 42, 43, 72; pyramids, 39–42; seasons, 58, 59–60; star tables and decans, 43, 62–4, 71–2
Ancient Greece: after-life, 110–11; cosmological models, 128; elements and properties of matter, 89–90; gods, 111; Hippocrates, 93–4; medicine, 93–4; and meteorites, 167; theory of the psyche, 90–1
Ancient Greek philosophers, 89–91, 108, 128, 176; *see also* Plato; Pythagoras
Ancient Orient, 117
Anders, William, 255, 256–7, 260
animal sacrifice, 26, 27
animals in space, 218, 236–8, *237*, 254–5
animism, 23–4, 167
Apollo missions: Apollo 8, 255–7; Apollo 9, 262; Apollo 11, 245–6, 264; Apollo 13, 247; Apollo 14, 241, 263; Apollo 17, 260, 261
Aquarius, age of, 231–2, 246
Aratus, *Phaenomena*, 70
archetypes (in psychology), 233–5
Arcturus, 74, 77
Arendt, Hannah, *The Human Condition*, 243–5
Aries, 94
Aristarchus of Samos, 128
Aristotle, 114–15, 116, 276n49
armillary spheres, 144–5; *see also* astronomical instruments (pre-telescope)
Armstrong, Neil, 245, 246
art: prehistoric, 8, 16–18; space exploration depicted in, 216–17
astral plane, 225, 227